Soft Computing for
Intelligent Robotic Systems

Studies in Fuzziness and Soft Computing

Editor-in-chief
Prof. Janusz Kacprzyk
Systems Research Institute
Polish Academy of Sciences
ul. Newelska 6
01-447 Warsaw, Poland
E-mail: kacprzyk@ibspan.waw.pl

Lakhmi C. Jain
Toshio Fukuda (Eds.)

Soft Computing for Intelligent Robotic Systems

With 131 Figures
and 21 Tables

Springer-Verlag
Berlin Heidelberg GmbH

Prof. Lakhmi C. Jain
Knowledge-Based Intelligent Engineering Systems Centre
University of South Australia
Adelaide, Mawson Lakes, S.A., 5095, Australia

Prof. Toshio Fukuda
Nagoya University
Center for Cooperative Research in
Advanced Science and Technology
Furo-cho, Chikusa-ku
Nagoya 464-01, Japan

ISBN 978-3-662-13003-2

Cataloging-in-Publication Data applied for
Die Deutsche Bibliothek – CIP-Einheitsaufnahme
Soft computing for intelligent robotic systems; with 21 tables / Lakhmi C. Jain; Toshio
Fukuda (eds.).
  (Studies in fuzziness and soft computing; Vol. 21)
  ISBN 978-3-662-13003-2   ISBN 978-3-7908-1882-6 (eBook)
  DOI 10.1007/978-3-7908-1882-6

© Springer-Verlag Berlin Heidelberg 1998
Originally published by Physica-Verlag Heidelberg New York in 1998
Softcover reprint of the hardcover 1st edition 1998

The use of general descriptive names, registered names, trademarks, etc. in this publica-
tion does not imply, even in the absence of a specific statement, that such names are
exempt from the relevant protective laws and regulations and therefore free for general
use.

Hardcover Design: Erich Kirchner, Heidelberg

SPIN 10689238   88/2202-5 4 3 2 1 0 – Printed on acid-free paper

PREFACE

The tremendous growth in the use of soft computing techniques in the last few years can be attributed to the advances in hardware, software and sensor technology tools. Today, we have at our disposal these very powerful tools for building intelligent systems. The main purpose of this book is to report research results from some of the more recent applications of soft computing techniques for intelligent robotics systems.

It is not easy to present a full review of the field in any manageable book. Thus, our approach is to include the most recent advances of the soft computing techniques in robotics.

This book comprises nine chapters. The first chapter written by Watanabe and Izumi discusses a fuzzy-neural realization of behavior-based control systems for a mobile robot. Neural network, fuzzy system and genetic algorithms are used to control a mobile robot. The obstacle avoidance problem is used to demonstrate the feasibility of this research.

Chapter 2, by Ishiguro, Watanabe, Kondo and Uchikawa explores artificial immune networks and their applications to robotics. A new decentralized consensus-making mechanism based on the biological immune system is proposed. This scheme is validated by applying it to behavior arbitration for an autonomous mobile robot.

Chapter 3, by Johannet and Sarda presents reinforcement learning of a six-legged robot to walk and avoid obstacles. This research proposes a neural network based system which allows a small six-legged robot to walk and avoid obstacles even when partially damaged.

In Chapter 4, Urban, Buessler, and Gresser introduce an application of neural network techniques to robotic control. Arm movements are controlled by using visual features. The neurocontroller adapts on-line without any prior knowledge of the system geometry.

In Chapter 5, Kim and Lewis present an intelligent optimal design of CMAC neural network for robot manipulations. A new hierarchical intelligent control scheme using neural networks is proposed for a robotic manipulation.

Chapter 6, by Hasegawa and Fukuda proposes a new controller and a learning algorithm. The proposed controller is applied to the problem of controlling a seven-link brachiation robot, which moves dynamically from branch to branch like a gibbon, a long-armed ape, swinging its body in a pendulum.

Chapter 7, by Sehad discusses neural reinforcement learning for robot navigation. Sehad has shown that the Kohonen implementation of Q-learning is more suitable for real world applications than previously reported methods. Experiments in learning obstacle-avoidance behavior with the Khepera robot illustrate the efficiency of the proposed approach.

Chapter 8, by Ahrns, Bruske, Hailu, and Sommer, is on neural fuzzy techniques in sonar-based collision avoidance. The application of neuro-fuzzy control to collision avoidance by the TRC labmate robot is presented in this chapter.

In the last chapter, Meeden and Kumar present a review on the use of evolutionary computing techniques for the automatic design of adaptive robots. The focus is on methods that use neural networks and have been tested on actual robots.

This book will be useful for application engineers, scientists and researchers who wish to use soft computing techniques in robotics systems.

We are grateful to all of the contributors to this book. We would like to express our sincere thanks to Berend Jan van Zwaag for his excellent help in the preparation of the manuscript. We thank the Editorial Staff of Springer Verlag Company for their support. We would also like to thank Professor Janusz Kacprzyk for the opportunity to publish this book.

L.C. Jain, Australia
T. Fukuda, Japan

Contents

X

Chapter 1

A FUZZY-NEURAL REALIZATION OF BEHAVIOR-BASED CONTROL SYSTEMS FOR A MOBILE ROBOT

Keigo Watanabe and **Kiyotaka Izumi**

Department of Mechanical Engineering
Faculty of Science and Engineering
Saga University
1-Honjomachi, Saga 840, Japan
Tel: +81-952-28-8602 Fax: +81-952-28-8587
E-mail: watanabe@me.saga-u.ac.jp

A fuzzy-neural realization of a behavior-based control system is described for a mobile robot by applying the soft-computing techniques, in which a simple fuzzy reasoning is assigned to one elemental behavior consisting of a single input-output relation, and then two consequent results from two behavioral groups are competed or cooperated. For the competition or cooperation between behavioral groups or elemental behaviors, a suppression unit is constructed as a neural network by using a sign function or saturation function. A Jacobian net is introduced to transform the results obtained from the competition or cooperation to those in the joint coordinate systems. Furthermore, we explain how to learn the present behavior-based control system by using a genetic algorithm. Finally, a simple terminal control problem is illustrated for a mobile robot with two independent driving wheels.

1 Introduction

In the control of mobile robots and manipulators, behavior-based control has been actively studied [1]-[6], whose control system is intelligently constructed by running the robot in a given complex environment. This approach is different from conventional approaches such as a regulator problem and a servo problem in the construction of control system. As a representative, there exists the so-called subsumption architecture [7]. The main features of this concept are that the control system is divided into behavioral elements (or modules), instead of dividing it into functional elements, and the behavioral fusion is taken through the competition and cooperation between behavioral elements. That is, in the subsumption architecture, control is distributed among task-achieving behaviors that operate asynchronously, only one behavior actually controlling the robot at a time. From this point of view, the subsumption architecture can also be interpreted as one example of parallel processing architectures or of bottom-up approaches to building a control system [8].

There have been already several studies of subsumption architecture-like control systems, depending on how to realize such a fundamental concept. Although a fusion method between behavioral elements was fixed in advance for the conventional subsumption architecture, Maes [9] proposed a behavior net approach, in which all behavioral elements are connected by the network and the causality relation between behavioral nodes is learned. Oka *et al.* [10] studied an extended approach that combined the Maes behavior net with conventional feedforward and feedback control methods. Since these approaches need the so-called regional knowledge, it is necessary that the detailed behavioral knowledge and strategy are provided in advance.

On the contrary, there exits the Koza GP (Genetic Programming) approach [11], as a method which acquires the behavioral knowledge from actual behaviors, instead of providing a priori regional knowledge. A dynamic behavioral knowledge is evolutionarily obtained through this approach. Note, however, that it needs more number of individuals and generations, and the resultant rules do not seem to be robust against behaviors, because its rules are not generated from fuzzy reasoning techniques.

To acquire the learning behavior of obstacle avoidance control problem for a mobile robot, Furuhashi *et al.* [12] had studied a GA (Genetic Algorithm) approach similar to the Pittsburgh approach that is well-known as a machine learning, in which all control rules are represented using one individual. However, in this approach, the fuzzy reasoning mechanism is constructed using the fusion of several sensor information as conventional control, so that it does not seem to be a behavior-based control or a bottom-up approach, even if it is regarded as a learning-type control, because a reasoning is not generated using the behavioral element and there are no competition and cooperation between behavioral elements.

In this chapter, a fuzzy behavior-based control [13]-[15] is described to build a control system by using a bottom-up approach, in which a simple fuzzy reasoning is assigned to one behavioral element consisting of single-input and single-output and the competition and cooperation of fuzzy reasoning consequent between behavioral groups are introduced. Here, it was assumed that, for the simplicity of problem, the number of input and output order is same for any behavioral group and it is also same between behavioral groups

In what follows, we first explain the input-output relation of any behavioral group and its fuzzy rules within such a behavioral group. Next, the fusion method is described by using the competition and cooperation between behavioral groups or elements. In particular, a suppression unit is proposed as a neural network using a sign or saturation function. We also introduce a Jacobian net to transform the fuzzy reasoning results obtained from the competition and cooperation into the quantities in the motor (or joint) coordinates. Next, a construction example is discussed for a simple terminal control problem of a mobile robot in an environment with an obstacle. Moreover, a conventional GA is applied to acquire some suitable fuzzy-based behavioral groups for a given task. Finally, simulation examples are demonstrated through an obstacle avoidance problem for a mobile robot with two independent driving wheels.

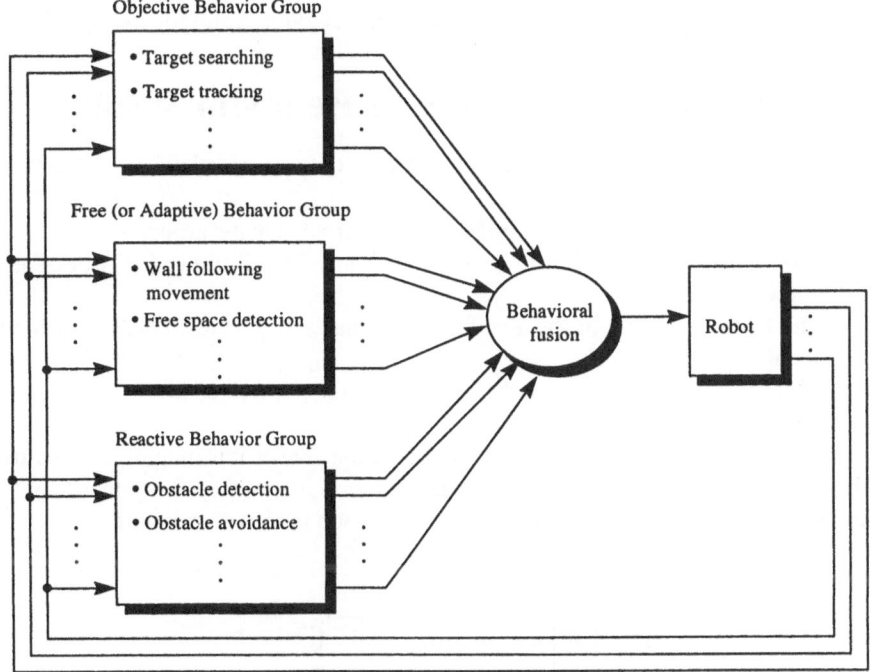

Figure 1. A Behavior-Based Control System Using Multiple Behavior Groups for a Mobile Robot

2   Input-Output Relation of Behavior Group and Its Fuzzy Rules

2.1 Behavior-Based Control Systems

Figure 1 shows an example of the behavior-based control system consisting of three behavior groups for a mobile robot. The reactive behavior group basically consists of some behavioral elements that detect or avoid obstacles. The free (or adaptive) behavior group is composed of some behavioral elements that move following the wall or detect a free space; these behavioral elements connect with a higher level behavior. The objective behavior group consists of some behavioral elements that search or track a target, which is an essential purpose (or task) of the robot. The behavioral fusion (unification) part fuses the behavioral results generated from each behavior group by using the competition and cooperation.

It should be noted that this figure presents a basic concept for the behavior-based control system, so that we have to consider how to use the sensor information obtained from the robot and how to construct each behavior group for each problem.

2.2 Input-Output Relation of Behavior Group

In the implementation of a behavior group, the following assumptions are made:

- The number of input-output order is same for a behavior group.
- The sensor information is generated from the absolute (or work-space) coordinates.
- The force or torque in the absolute coordinates is generated from a behavior group.
- Other behavior groups have the same conditions described above.

As an additional assumption, it is assumed that if the sensor information is once used in a behavior group, then the same sensor information is never used in other behavior group. The input-output relation of this behavior group is shown in Figure 2.

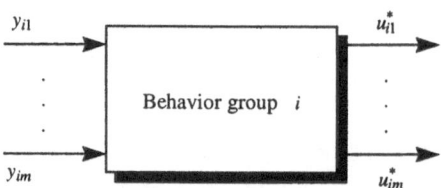

Figure 2. Input-Output Relation for a Behavior Group

2.3 Fuzzy Rule Relation of Behavior Group

As can be seen from Figure 3, in a behavior group, a simple fuzzy reasoning is applied to one behavioral element using one sensor information y, and it generates one reasoning result u^*. We use a Gaussian-type function as the membership function, and use the simplified reasoning [16]. The resultant fuzzy reasoning consequent is obtained as

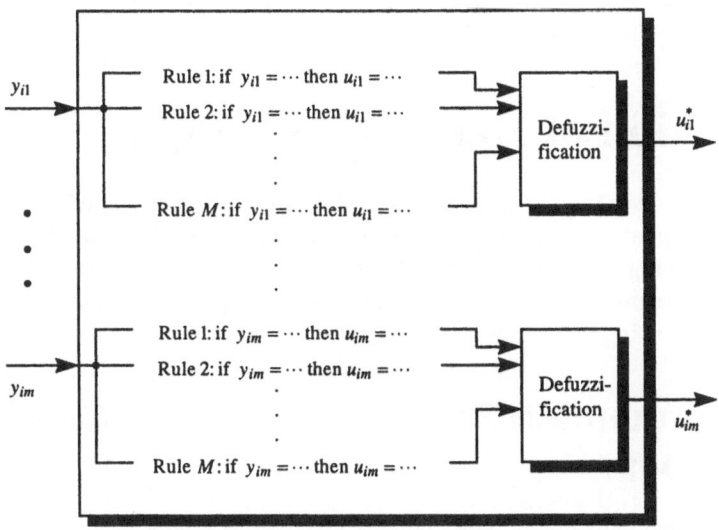

Figure 3. Internal Fuzzy Rules for a Behavior Group i

$$u^* = \sum_{i=1}^{M} p_i w_{bi} \tag{1}$$

where M denotes the total number of rules, w_{bi} the constant in the conclusion of the i-th rule, and p_i the normalized rule confidence such as

$$p_i = \frac{\mu_i}{\sum_{j=1}^{M} \mu_j} \tag{2}$$

$$\mu_i = \exp\{\ln(0.5)(y - w_{ci})^2 w_{di}^2\} \tag{3}$$

Here, w_{ci} denotes the center value (e.g. the mean value of a Gaussian like membership function) associated with the i-th membership function, and w_{di} denotes the reciprocal value of the deviation from the center w_{ci} to which the i-th Gaussian function of the input data on the support set has value 0.5.

In a conventional reactive behavior, a number of rules are driven in parallel by using only one sensor information and a suitable rule that satisfies a selected condition. On the other hand, using a simple fuzzy reasoning technique gives a consequent result as a weighted sum of multiple rules. In other words, using M fuzzy divisions of a behavioral element gives a tolerance to the determination of the behavioral element for an environment; as a result, our approach may acquire a more robust behavior against a change of the environment.

3   Fusion of Behavior Groups

In this section, we explain how to fuse the reasoning results obtained from each behavioral element. That is, the forces or torques in the absolute coordinate systems are fused by using the competition and cooperation of the reasoning consequent for the behavioral elements between the behavior groups, and the associated fusion units are constructed using the neural network (NN).

3.1 Subsumption Relation of Behavior Group Outputs

Consider the two behavior groups i and $i+1$ shown in Figure 4. Let the two forces (or torques), which have the same output number in their behavior groups, be described as a and b, and their signs be compared. If they have the same sign, then it is regarded as the cooperation and the output from the upper group $i+1$ is treated as the fusion result $c = b$. Otherwise, if they have the opposite sign, then it is regarded as the competition and the output from the lower group i is treated as the fusion result $c = a$.

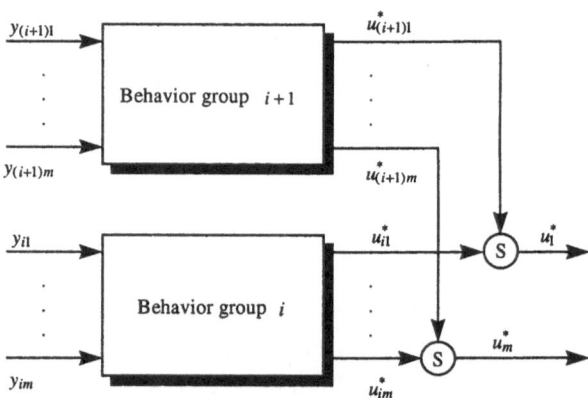

Figure 4. Fusion of Two Behavior Groups

In case of more than three groups, the similar fusion procedure is taken from the top group and proceeded to the lower group; thus, the force or torque that is fed to the Jacobian net is finally determined in the absolute coordinate systems. It should be noted that the fusion method described above has not necessarily been applied to the conventional subsumption architecture. For example, let a and b be the lower and the upper behaviors respectively, if either a or b has the OFF state, then the either ON state is generated as the fusion result, whereas if both a and b have the ON state, then the ON state in the lower behavior is generated as the fusion result.

3.2 Construction of Suppression Unit as a Neural Network

We refer to the unit with the logic described above as suppression unit. If we use the following sign function

$$\text{sgn}(x) = \begin{cases} 1 \text{ for } x > 0 \\ -1 \text{ for } x < 0 \end{cases} \tag{4}$$

the cooperation taking the same sign is written by

$$\text{if } |\text{sgn}(a) + \text{sgn}(b)| = 2 \text{ then } c = b \tag{5}$$

whereas the competition taking opposite sigh is written by

$$\text{if } |\text{sgn}(a) + \text{sgn}(b)| = 0 \text{ then } c = a \tag{6}$$

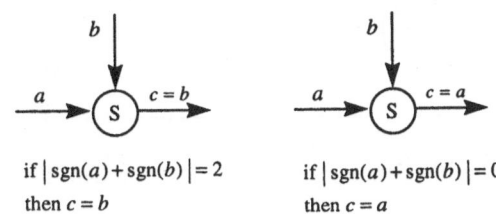

if $|\text{sgn}(a) + \text{sgn}(b)| = 2$ if $|\text{sgn}(a) + \text{sgn}(b)| = 0$
then $c = b$ then $c = a$

Figure 5. Logic for a Suppression

Figure 5 depicts this concept. Introducing a switching variable s and summarizing the above results gives

$$c = (1 - s)a + sb \tag{7}$$

$$s = |\text{sgn}(a) + \text{sgn}(b)| / 2 \tag{8}$$

This suppression logic circuit is shown in Figure 6, using a NN.

Note here that in this figure the unit with the symbol 1 generates the output of 1, the unit with the symbol Σ outputs the summation of the inputs. Similarly, the unit with the symbol Π outputs the product of the inputs. Furthermore, the unit with no symbols simply distributes the input to the output, and the unit that has symbols Σ and h generates the output through the absolute function with a linear summed input.

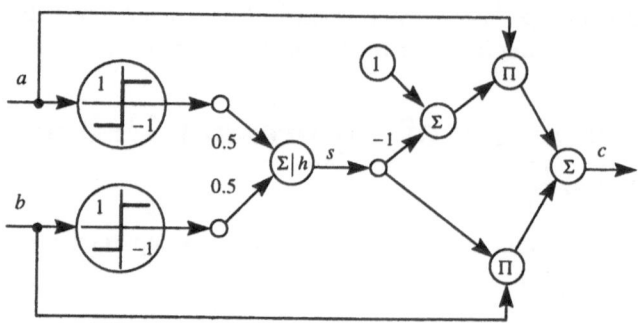

Figure 6. Sign Function-Type Suppression Unit Realized as a NN

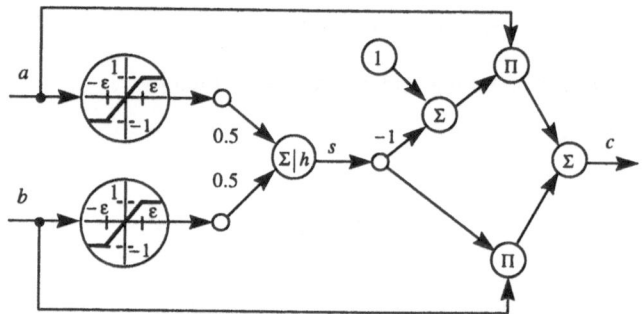

Figure 7. Saturation Function-Type Suppression Unit Realized as a NN

This sign function (or hard limiter) type suppression unit may not give a practical fusion result, if the output from both groups or either group is very small. Therefore, it seems to be practical that the above logic is applied to the reasoning results a and b having the larger magnitude than a fixed order, otherwise the fusion unit is composed of the weighted sum of both reasoning results. For this purpose, it is reasonable to introduce the following saturation function:

$$\text{sat}(x) = \begin{cases} \text{sgn}(x) \text{ for } |x| > \varepsilon \\ x/\varepsilon \text{ for } |x| \le \varepsilon \end{cases} \tag{9}$$

where ε denotes a small positive number. Consequently, following the same procedure as in (7), the output expression of the unit results in

$$c = (1-s)a + sb \tag{10}$$

$$s = |\text{sat}(a) + \text{sat}(b)| \, /2 \tag{11}$$

Figure 7 shows the suppression logic circuit based on using the saturation function, which is realized by using the NN.

3.3 Jacobian Net

The force or torque in the absolute (or work-space) coordinate systems is obtained by fusing the reasoning results generated from each behavior group by using the nonlinear suppression unit. To transform the above quantity to the torque in the motor (or joint) coordinate systems, we use the Jacobian net as shown in Figure 8. If the Jacobian matrix J is known, then the net structure is fixed as J^T; this is a simple coordinate transformation. In a practical robot operation, we have to further transform the torque input to the associated voltage or current, so that the transformation gain to the actuator driver should be determined in advance. However, for the case when the Jacobian matrix or transformation gain is unknown, the Jacobian net must be trained as a net with the changeable connection weights.

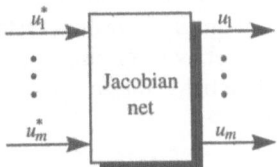

Figure 8. Jacobian Net

4   Construction Examples

Now, following the method described above, let us try to construct a fuzzy behavior-based control system of a terminal control problem for a mobile robot with two independent driving wheels.

4.1 Behavior Model for a Mobile Robot

Suppose that the sensor information of the robot is given by Figure 9. Here, ϕ is the azimuth angle of the robot, v the forward velocity, ψ the relative angle between the forward direction of the robot and the obstacle, d the distance between the robot and the obstacle, Ψ the relative angle between the forward direction of the robot and the objective point, and D the distance between the robot and the objective point. The task given for the robot is that the robot travels to an objective point, but if an obstacle is found, then the robot has to avoid it.

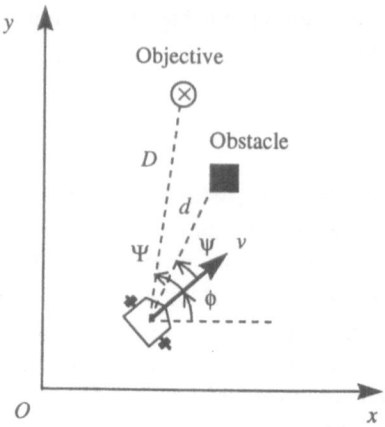

Figure 9. Behavior Model for a Mobile Robot with
Two Independent Driving Wheels

Figures 10 and 11 show two construction examples of the fuzzy behavior-based control system for the above problem. The former system (Case A) consists of an objective behavior group and a free behavior group, whereas the latter system consists of an objective behavior group, a free behavior group and a reactive behavior group.

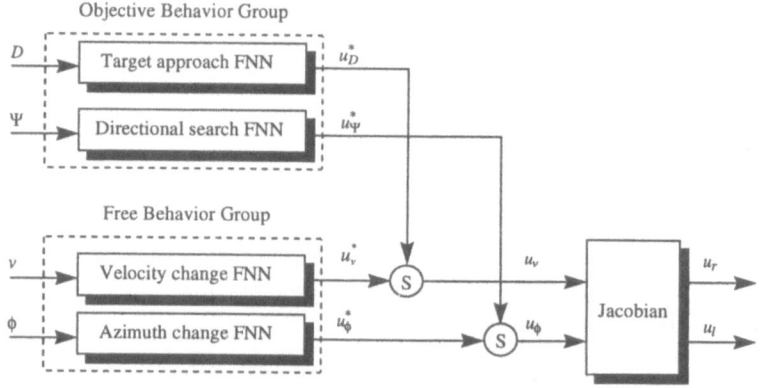

Figure 10. Fuzzy Behavior-Based Control System for a Mobile
Robot with Two Behavior Groups

Here, note that the sensor information supplied to each behavior group are not shared, as described before. In addition, note that in these figures u_D^* denotes the force to change the goal distance, u_Ψ^* is the torque to change the relative angle to the goal point, u_v^* is the force to change the velocity of the robot, and u_ϕ^* is the torque to change the azimuth of the robot. Furthermore, note that in Figure 11 u_d^* denotes the force to change the approaching distance to the obstacle and u_ψ^* denotes the torque to change the avoidance angle against the obstacle.

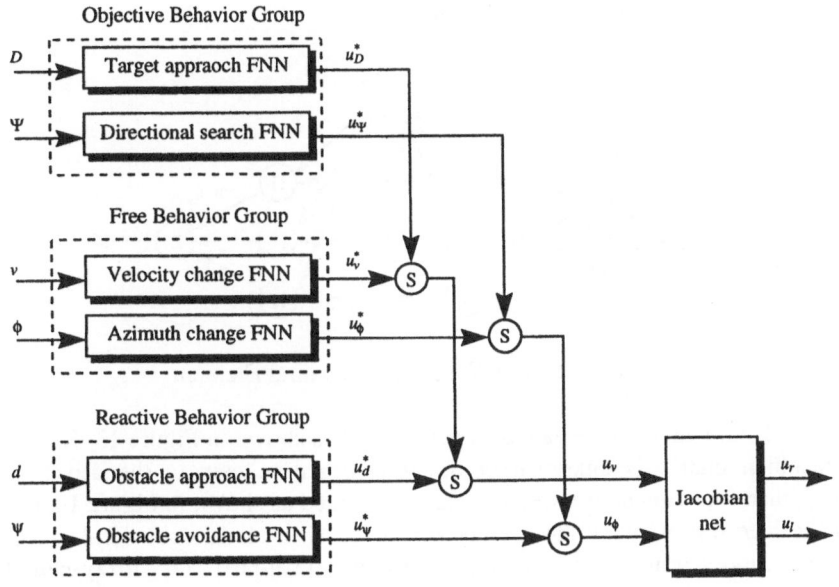

Figure 11. Fuzzy Behavior-Based Control System for a Mobile Robot
with Three Behavior Groups

4.2 Fuzzy-Neural Network Model for a Behavioral Element

Figure 12 shows one sample behavioral element, expressed by a fuzzy neural network (FNN), for the velocity change behavior in the free behavior group, in which the behavioral element is divided into three fuzzy regions. In this model, the center value w_{vc}, the reciprocal value of the deviation w_{vd}, and the constant value in the conclusion w_{vb} must be learned by using any learning method. Furthermore, the Jacobian matrix for this type of mobile robots is expressed as

$$\begin{bmatrix} u_r \\ u_l \end{bmatrix} = J^T \begin{bmatrix} u_v \\ u_\phi \end{bmatrix} \tag{12}$$

$$J^T = \frac{r}{2}\begin{bmatrix} 1 & 1/l \\ 1 & -1/l \end{bmatrix} \tag{13}$$

where u_r and u_l denote the driving torques for the right and left wheels, u_v is the force for changing the robot velocity, u_ϕ denotes the torque for changing the azimuth angle of the robot, r is the radius of the wheel, and l is the length from the center of gravity (c.g.) of the robot to the one wheel.

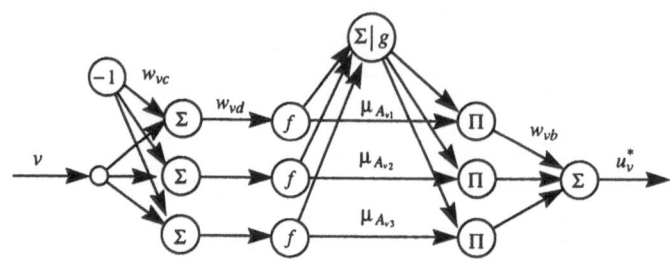

Figure 12. An FNN for a Behavioral Element

Thus, if J is known, then the Jacobian net is fixed as shown in Figure 13. However, if the Jacobian matrix is unknown or the transformation gain to the driver is also learned, then four connection weights must be learned as shown in Figure 14, together with the learning of fuzzy parameters associated with all of behavioral groups, by applying any learning or training algorithms [16]-[18] for fuzzy-neural (or neurofuzzy) systems. In the following section, we will use a GA to learn the fuzzy reasoning parameters w_{vc}, w_{vd}, and w_{vd} in each behavioral group, assuming that the Jacobian matrix is completely known.

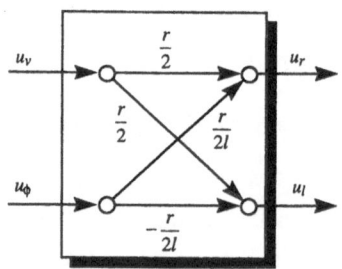

Figure 13. Jacobian Net: Known Case

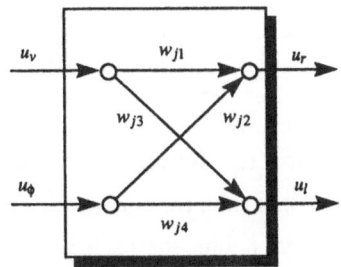

Figure 14. Jacobian Net: Unknown Case

5 Learning of Control System Using Genetic Algorithms

5.1 Fitness Function

The objective behavior is that the mobile robot travels from the point (0, 0) to the final point (x_f, y_f) in the absolute coordinate systems. Let the fitness of GA be given by

$$fitness = \begin{cases} D+p \text{ if } d(t) \le 0 \\ \quad D \text{ if } d(t) > 0 \end{cases} \tag{14}$$

with

$$p = 5\sum_{i=0}^{N} s_i \tag{15}$$

$$s_i = \begin{cases} 1 \text{ if } d(i\Delta T) \le 0 \\ 0 \text{ if } d(i\Delta T) > 0 \end{cases} \tag{16}$$

where N is the number of total control steps, i is the discrete time index, and ΔT is a sampling interval. The equation (14) means that a penalty p is added to the normal cost D, if the mobile robot would collide with the obstacle (or the mobile robot would be in the area of the obstacle). Acquiring the objective behavior is equivalent to solving the minimization problem of (14). The equation (14) for Case A is evaluated, assuming that the position information on the obstacle are given in advance without any sensor. On the other hand, the equation (14) for Case B is evaluated, assuming that the position and azimuth information on the obstacle are provided by two sensors.

5.2 Coding

It is assumed that the fuzzy reasoning for each behavioral element consists of a simplified reasoning [16] with single input and output and that a Gaussian type function described in (3) is used as the membership function. In order to learn this fuzzy reasoning with GA, the phenotype of one fuzzy rule is composed of the center value of the membership function α_{ij}, the reciprocal value of the standard deviation β_{ij} and the constant value of conclusion (or consequent) part B_{ij}, as shown in Figure 15. The genotype is coded with 8 bits for each parameter. Here, the subscript i denotes the index of behavioral element and j denotes the fuzzy rule number. The transformation from the genotype to the phenotype is given by

$$x = x_{\min} + (x_{\max} - x_{\min})\frac{gcode}{2^8 - 1} \tag{17}$$

where x is the center value of membership function or the reciprocal value of standard deviation or the constant value of conclusion, $x_{\min}$ and $x_{\max}$ denote the minimum and maximum values for each parameter, and $gcode$ denotes the value obtained by transforming the genotype into the integer value with a gray code.

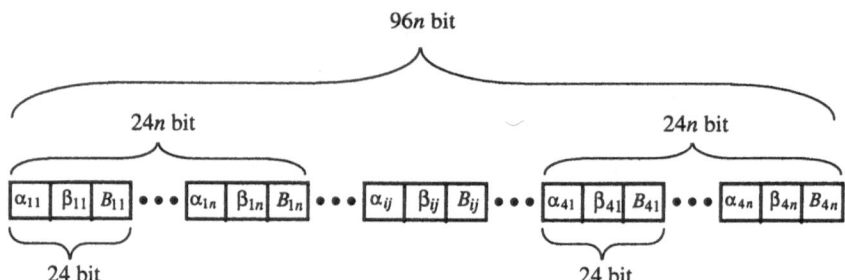

Figure 15. Individual Structure for Case A

If the number of labels is n, then an individual needs $n \times 8 \times 3$ bits for one behavioral element, because of the use of simplified fuzzy reasoning as described in (1). Accordingly, if the whole set of control rules is represented by one individual in a population, like Pittsburgh (or Pitts) approach [19], then the individual length needs $96n$ bits for Case A consisting of four behavioral elements, whereas it needs $144n$ bits for Case B consisting of six behavioral elements. See [20],[21] for the more detail GAs in a conventional fuzzy control problem.

6   Simulations

6.1 Mobile Robot Model

Let the mobile robot with two independent driving wheels be rigid moving on the plane. It is assumed that the absolute coordinate system *O-XY* is fixed on the plane as shown in Figure 16. Then, the dynamic property of the robot is given by the following equations of motion [18]:

$$I_v \ddot{\phi} = D_r l - D_l l \tag{18}$$

$$M \dot{v} = D_r + D_l \tag{19}$$

For the right- and left-wheels, the dynamic property of the driving system becomes

$$I_w \ddot{\theta}_i + c \dot{\theta}_i = k u_i - r D_i; \quad i = r, l \tag{20}$$

where each parameter and variable are defined by

I_v :   moment of inertia around the c.g. of robot
M :   mass of robot
D_l, D_r : left and right driving forces
l :   distance between left or right wheel and the c.g. of robot
ϕ :   azimuth of robot
v :   velocity of robot
I_w :   moment of inertia of wheel
c :   viscous friction factor
k :   driving gain factor
r :   radius of wheel
θ_i :   rotational angle of wheel
u_i :   driving input

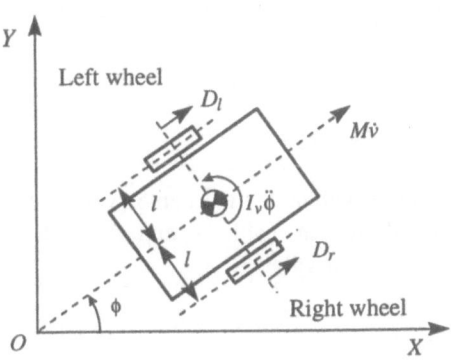

Figure 16. Model of a Mobile Robot with Two Independent Driving Wheels

The geometrical relationships among variables ϕ, v, θ_i are given by

$$r\dot{\theta}_r = v + l\dot{\phi} \tag{21}$$

$$r\dot{\theta}_l = v - l\dot{\phi} \tag{22}$$

From these equations, defining the state variable vector for the robot as $x(t) = [v(t)\ \phi(t)\ \dot{\phi}(t)]^T$, the manipulated variable vector as $u(t) = [u_r(t)\ u_l(t)]^T$, and the output variable vector as $y(t) = [v(t)\ \phi(t)\ ]^T$ yield the following state equation:

$$\dot{x}(t) = Ax(t) + Bu(t) \tag{23}$$

$$y(t) = Cx(t) \tag{24}$$

where

$$A = \begin{bmatrix} a_1 & 0 & 0 \\ 0 & 0 & 1 \\ 0 & 0 & a_2 \end{bmatrix}, \quad B = \begin{bmatrix} b_1 & b_1 \\ 0 & 0 \\ b_2 & -b_2 \end{bmatrix}, \quad C = \begin{bmatrix} 1 & 0 & 0 \\ 0 & 1 & 0 \end{bmatrix}$$

$$a_1 = -2c / (Mr^2 + 2I_w)$$

$$a_2 = -2cl^2 / (I_v r^2 + 2I_w l^2)$$

$$b_1 = kr / (Mr^2 + 2I_w)$$

$$b_2 = krl / (I_v r^2 + 2I_w l^2)$$

We considered a mobile robot that was constructed in our laboratory, whose physical parameters are given by

$$I_v = 0.6541 [\text{kgm}^2], \quad M = 25.5 [\text{kg}]$$

$$l = 0.165 [\text{m}], \quad r = 0.05 [\text{m}]$$

$$I_w = 0.4419 \times 10^{-3} [\text{kgm}^2], \quad k = 90$$

$$c = 0.0479 [\text{kgm}^2 / \text{s}]$$

It was assumed that the control sampling period was $\Delta T = 50 [\text{ms}]$, and the initial state was $x = [0\ 0\ 0]^T$.

6.2 Parameter Setting

The maximum and minimum values for the center value of membership function in the antecedent part are shown in Table 1 and those for the constant value of conclusion part are also shown in Table 2.

Table 1. Min and Max Values of Central Values

	Min	Max
v	−1	1
D	0	10
d	−1	5
ϕ, Ψ and ψ	−π	π

Table 2 Min and Max Values of Consequent Parts

	Min	Max
v, D and d	−0.5	0.5
ϕ, Ψ and ψ	−0.1	0.1

Note here that the maximum value of the reciprocal value of standard deviation was set to be 10 and the associated minimum value was set to be 0.1, where all behavioral elements were assumed to share those values and the number of labels was to be 5. The population size was taken to be 50. The selection method was a tournament type with three individuals. The crossover method was a uniform type with its probability 0.6. The mutation probability was fixed to be 1/480 in Case A and 1/720 in Case B.

When the generation that has a fitness less than equal 0.05 was successively appeared 3 times, the learning of GA was assumed to be stopped, because the fitness (14) becomes less than 0.05, if the objective has been achieved. Since the suppression logic circuit based on using the saturation function was applied to this example, the boundary layer value ε was taken to be 0.01.

6.3 Simulation Results for a Method with Two Behavioral Groups

Figures 17-22 illustrate the results of the path, speed, azimuth, input torques of the mobile robot, and the minimum fitness of each generation, where the individual that has satisfied the stopping condition described above is used to generate Figures 17-21, whose fitness is 0.0318 at the 40-th generation. Note here that the circle appeared at the right-hand upper side denotes the terminal area, whose center was located at $(x_f = 0.4, y_f = 0.4)$ with the radius 0.05 [m]. The circle at the center, drawn with a dot line, denotes the obstacle. It is found, from Figures 17 and 18, that the mobile robot has reached sufficiently to the final goal point, avoiding the obstacle. If a path that satisfies the cost of minimum distance or minimum fuel consumption is desired, it is necessary that an approach evaluation to the mid point or a constraint with respect to the energy consumption is included in the fitness function. Limiting to the mobile robot, if the maximum and minimum values of conclusion part shown in Table 2 are both to be smaller, then the torque u_ϕ for changing the azimuth can be reduced and therefore the unsteady response of the mobile robot can be suppressed. It is found from Figure 19 that the robot turned in the neighborhood of the final objective point, because the forward direction of the robot was suddenly changed after 18 seconds. It is also seen, from Figures 20 and 21, that the control inputs are slightly fluctuated, especially at the transient and ending stages. From Figure 22, observe that a fluctuating tendency of fitness is obtained to converge within 0.05.

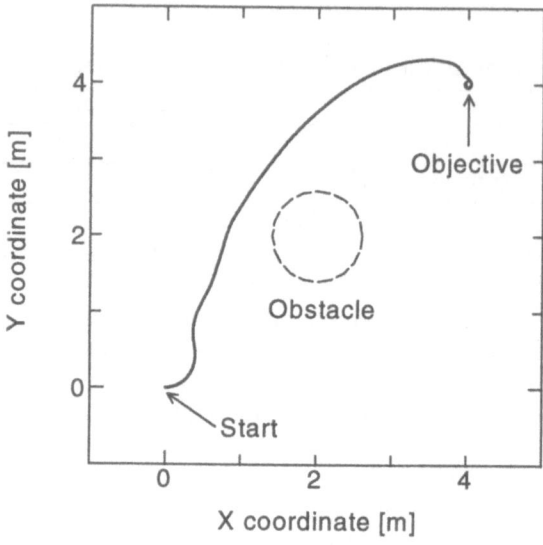

Figure 17. Trajectory of a Mobile Robot for Case A

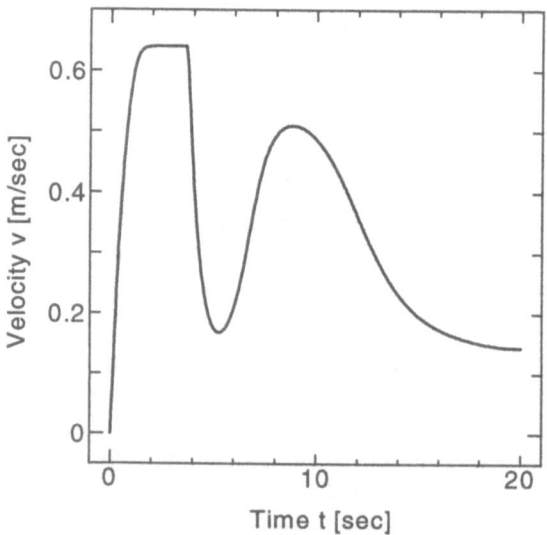

Figure 18. Velocity v Response for Case A

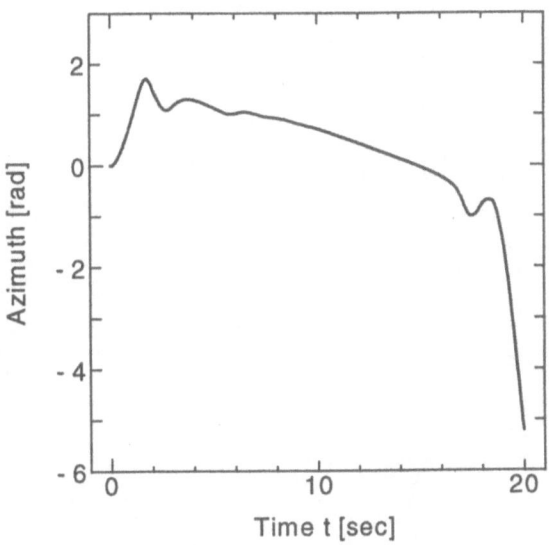

Figure 19. Azimuth φ Response for Case A

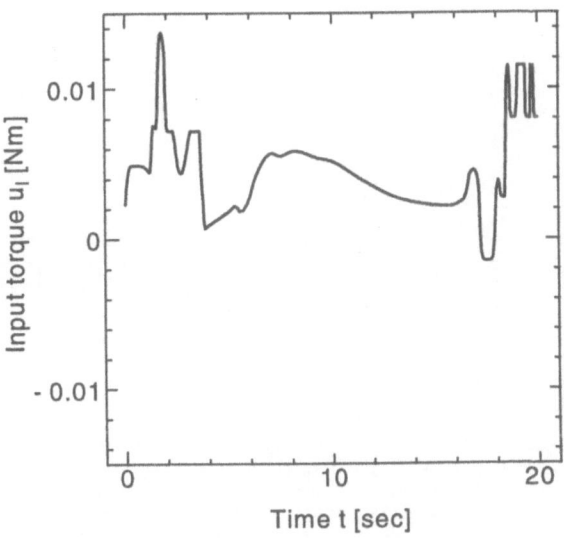

Figure 20. Control Input u_l for Case A

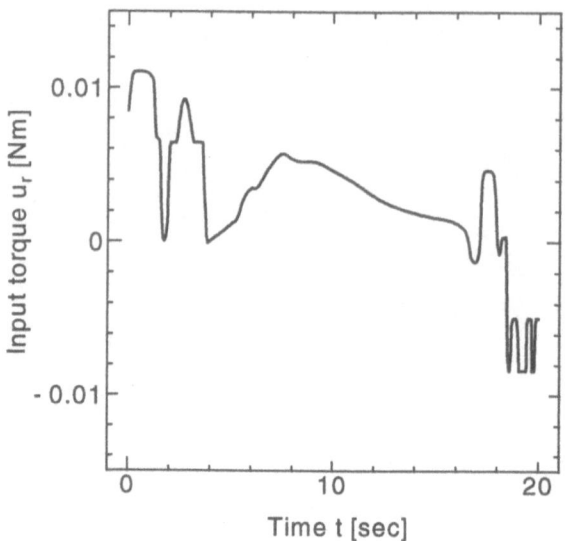

Figure 21. Control Input u_r for Case A

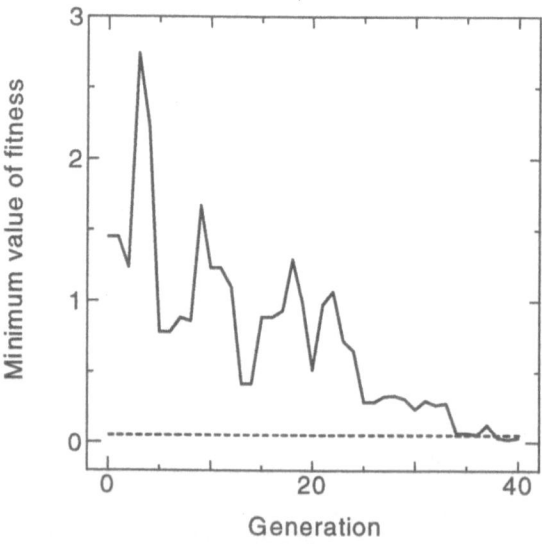

Figure 22. History of Minimum Value of Fitness for Case A

6.4 Simulation Results for a Method with Three Behavioral Groups

Next, Figures 23-28 illustrate the similar results for Case B, in which the final minimum fitness is 0.0285 at the 19-th generation. It should be noted that the robot in Figure 23 avoids the obstacle with a farther distance, compared with that in Figure 17. The bigger control inputs are obtained for this case, compared with Case A, so that the mobile robot has reached to the goal point with an earlier time than Case A. It should be noted, however, that after reaching the objective point the mobile robot successively turns until the final time, as can be seen from Figure 23. It is found from Figure 24 that the mobile robot travels with an approximate constant speed. However, note that the driving input torques of the right and left wheels give a considerable fluctuation, compared with those of Case A. Finally, observe that a convergence trend which is similar to that of Figure 22 is obtained in Figure 28.

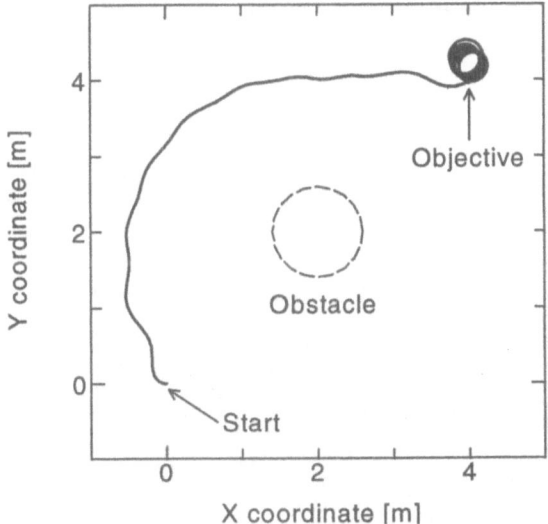

Figure 23. Trajectory of a Mobile Robot for Case B

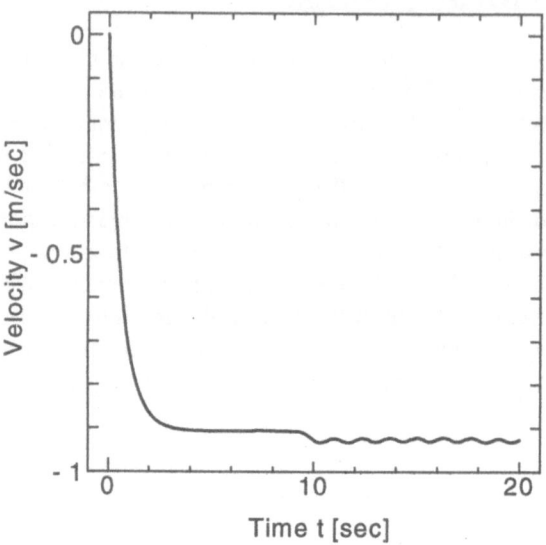

Figure 24. Velocity *v* Response for Case B

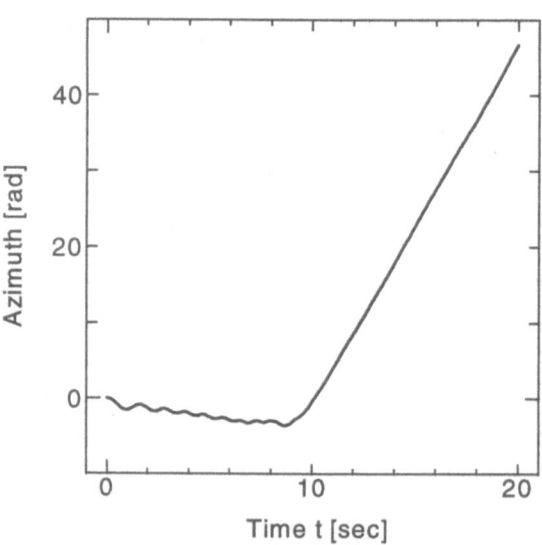

Figure 25. Azimuth φ Response for Case B

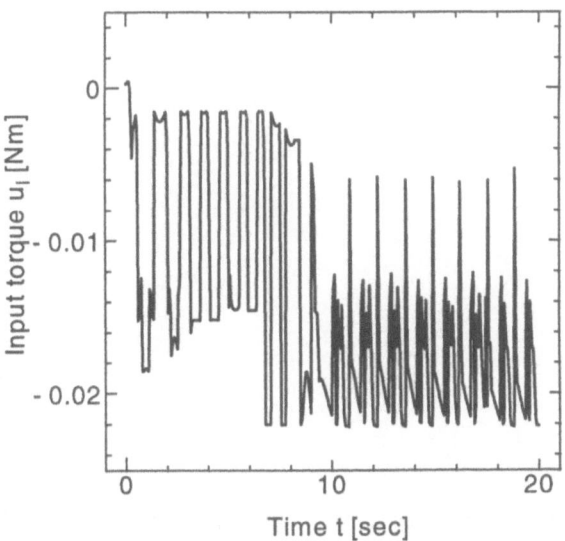

Figure 26. Control Input u_l for Case B

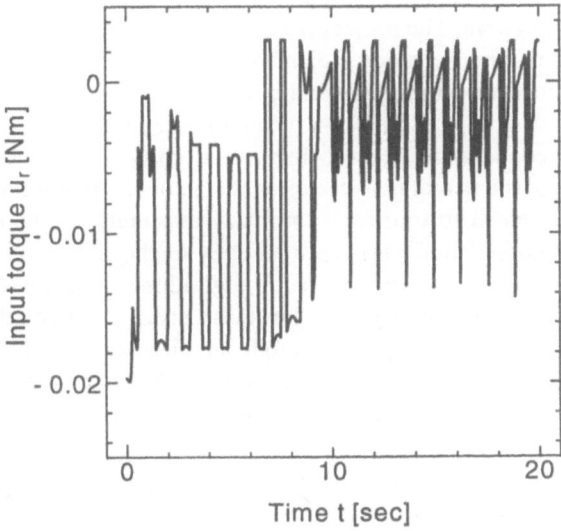

Figure 27. Control Input u_r for Case B

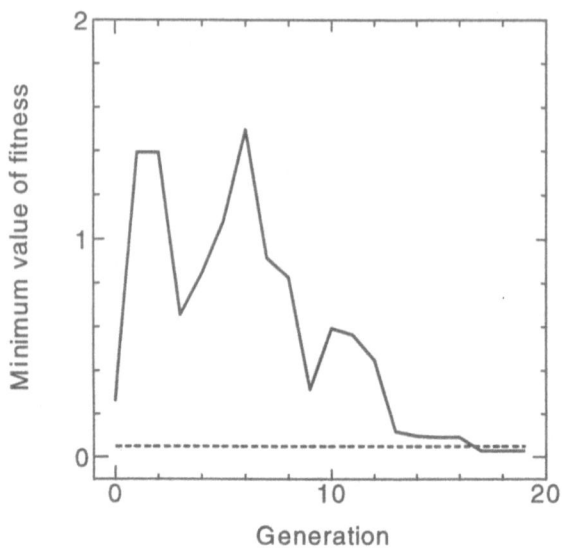

Figure 28. History of Minimum Value of Fitness for Case B

7   Conclusions

In this chapter, we have described a fuzzy behavior-based control, which is basically realized by the soft-computing techniques, as one of bottom-up approaches to building a control system. The present method has been based on introducing a simple fuzzy reasoning to one behavioral element consisting of single-input and single-output, and making the competition and cooperation of fuzzy reasoning consequent between behavioral groups. To make a fusion of fuzzy reasoning results by using the competition and cooperation between behavioral groups or elements, a suppression unit was described as a neural network using a sign or saturation function. A Jacobian net was also introduced to transform the fuzzy reasoning results obtained from the competition and cooperation into the quantities in the motor (or joint) coordinate systems. Moreover, a conventional genetic algorithm was applied to acquire some suitable fuzzy-based behavioral groups for a task given in an environment. Finally, a simple example was demonstrated through an obstacle avoidance problem for a mobile robot with two independent driving wheels. We are now planning to apply the present method to an omnidirectional mobile robot [22] and a practical miniature robot [23].

References and Further Reading

[1] S. Yamada, (1995), "Learning for Reactive Planning," *J. of the Robotics Society of Japan*, 13, 1, pp.38-43 (in Japanese).

[2] Y. Kuno, (1993), "Behavior of Behavior-based Robots," *J. of the Robotics Society of Japan*, 11, 8, pp.1178-1184 (in Japanese).

[3] N. Yoshida, (1993), "Robot Motion Control Using Subsumption Architecture," *J. of the Robotics Society of Japan*, 11, 8, pp.1118-1123 (in Japanese).

[4] A. Kara, K. Kawamura, S. Bagchi, and M. El-Gamal, (1992), "Reflex Control of a Robotic Aid System to Assist the Physically Disabled," *Control Systems Magazine*, 12, 3, pp.71-77.

[5] F. Mondada and E. Franzi, (1993), "Biologically Inspired Mobile Robot Control Algorithms," *Procs. of the NRP23--Symposium on Artificial Intelligence and Robotics*, Zurich, Switzerland.

[6] S. Nakasuka, T. Yairi, and H. Wajima, (1996), "Autonomous generation of reflex-based robot controller using inductive learning," *Robotics and Autonomous Systems*, 17, pp.287-305.

[7] R.A. Brooks, (1986), "A Robust Layered Control Systems for a Mobile Robot," *IEEE Robotics and Automation*, 2, 1, pp.14-23.

[8] H. Hu and M. Brady, (1996), "A parallel processing architecture for sensor-based control of intelligent mobile robot," *Robotics and Autonomous Systems*, 17, pp.235-257.

[9] P. Maes., (1992), "Learning Behavior Networks from Experience," *Procs. of the first European Conference on Artificial Life (ECAL-91)*, edited by F. J. Varela and P. Bourgine, pp.48-57

[10] T. Oka, M. Inaba, and H. Inoue, (1996), "Designing Hierarchical Motion Systems for Autonomous Robots based on BeNet," *Procs. of JSME Annual Conf. on Robotics and Mechatronics*, B, pp.1361-1364 (in Japanese).

[11] J.R. Koza, (1992), "Evolution of Subsumption using Genetic Programming," *Procs. of the first European Conference on Artificial Life (ECAL-91)*, edited by F. J. Varela and P. Bourgine, pp.110-119.

[12] T. Furuhashi, K. Nakaoka, H. Maeda, and Y. Uchikawa, (1995), "A Proposal of Genetic Algorithms with a Local Improvement Mechanism and Finding of Fuzzy Rules," *J. of Japan Society for Fuzzy Theory and Systems*, 7, 5, pp.978-987 (in Japanese).

[13] K. Watanabe and K. Izumi, (1997), "Construction of Fuzzy Behavior-Based Control Systems," *Reports of the Faculty of Science and Engineering*, Saga University, 25, 2, January, pp.1-7 (in Japanese).

[14] K. Izumi and K. Watanabe, (1997), "A Fuzzy Behavior-Based Control for an Autonomous Mobile Robot," *Reports of the Faculty of Science and Engineering*, Saga University, 25, 2, January, pp.167-175 (in Japanese).

[15] K. Izumi and K. Watanabe, (1997), "Fuzzy Behavior-Based Tracking Control for a Mobile Robot," *Procs. of 2nd Asian Control Conference*, Seoul, Korea, 1, pp.685-688

[16] K. Watanabe, K. Hara, S. Koga, and S.G. Tzafestas, (1995), "Fuzzy-neural network controllers using mean-value-based functional reasoning," *Neurocomputing*, 9, pp. 39-61.

[17] C.J. Harris, M. Brown, K.M. Bossley, D.J. Mills, and F. Ming, (1996), "Advances in Neurofuzzy Algorithms for Real-time Modelling and Control," *Engng Applic. Artificial Intelligence*, 9, 1, pp.1-16.

[18] K. Watanabe, J. Tang, M. Nakamura, S. Koga, and T. Fukuda, (1996), "A Fuzzy-Gaussian Neural Network and Its Application to Mobile Robot Control," *IEEE Trans. on Control Systems Technology*, 4, 2, pp.193-199.

[19] Z. Michalewicz, (1996), *Genetic Algorithms + Data Structures = Evolution Programs*, Springer.

[20] D.A. Linkens and H.O. Nyongesa, (1995), "Genetic algorithms for fuzzy control Part 1: Offline system development and application," *IEE Proc.-Control Theory Appli.*, 142, 3, pp.161-176.

[21] D.A. Linkens and H.O. Nyongesa, (1995), "Genetic algorithms for fuzzy control Part 2: Online system development and application," *IEE Proc.-Control Theory Appli.*, 142, 3, pp.177-185.

[22] J. Tang, K. Watanabe, and Y. Shiraishi, (1996), "Design of Traveling Experiment of an Omnidirectional Holonomic Mobile Robot," *Procs. of 1996 IEEE/RSJ Int. Conf. on Intelligent Robotics and Systems (IROS96)*, Osaka, Japan, 1, pp.66-73.

[23] F. Mondada, E. Franzi, and P. Ienne, (1993), "Mobile robot miniaturisation: A tool for investigation in control algorithms," *Procs. of the 3rd Int. Symposium on Experimental Robotics*, Kyoto, Japan, October 28-30.

Chapter 2

ARTIFICIAL IMMUNE NETWORK AND ITS APPLICATION TO ROBOTICS

Akio Ishiguro, Yuji Watanabe, Toshiyuki Kondo and **Yoshiki Uchikawa**
Department of Computational Science and Engineering
Graduate School of Engineering
Nagoya University
Furo-cho, Chikusa-ku, Nagoya 464-01
Japan
e-mail: {ishiguro/yuji/kon/uchikawa}@bioele.nuee.nagoya-u.ac.jp

Conventional Artificial Intelligence (AI) techniques have been criticized for their brittleness under dynamically changing environments. In recent years, therefore, much attention has been focused on the reactive planning approach such as behavior-based AI. However, in the behavior-based artificial AI approach, there are following problems that have to be resolved: 1) how do we construct an appropriate arbitration mechanism, and 2) how do we prepare appropriate behavior primitives (competence modules). On the other hand, biological information processing systems have various interesting characteristics viewed from the engineering standpoint. Among them, in this study, we particularly pay close attention to the immune system. We try to construct a decentralized consensus-making mechanism inspired by the immune network hypothesis. To tackle the above-mentioned problems in the behavior-based AI, we apply the proposed method to behavior arbitration for an autonomous mobile robot by carrying out some simulations and experiments using a real robot. In addition, we investigate two types of adaptation mechanisms to construct an appropriate artificial immune network without human intervention.

1 Introduction

In recent years much attention has been focused on behavior-based AI, which has already demonstrated its robustness and flexibility against dynamically changing world. In this approach, intelligence is expected to result from both mutual interactions among behavior primitives (competence modules) and interaction between the robot and environment. However, there are still open questions:

- How do we construct a mechanism that realizes appropriate arbitration among multiple competence modules?
- How do we prepare appropriate competence modules?

Brooks recently showed a solution to the former problem with the use of the *subsumption architecture* [1][2]. Although this method demonstrates highly robustness, it should be noted that this architecture arbitrates the prepared competence modules on a *fixed priority* basis. It would be quite natural to vary the priorities of the prepared competence modules according to the situation.

Maes proposed an another flexible mechanism called the *behavior network system* [3][4]. In this method, agents (i.e. competence modules) form a network based on their cause-effect relationship, and an agent suitable for the current situation and the given goals emerges as the result of activation propagation among agents. This method, however, is difficult to apply to a problem where it is hard to find the cause-effect relationship among agents.

One of the promising alternative approaches to tackle the above-mentioned problems is a biologically inspired approach. Among the biological systems, we particularly focus on the immune system, since this system is dedicated to self-preservation under hostile environment (needless to say autonomous mobile robots must cope with dynamically changing environment). This system additionally has various interesting features such as *immunological memory*, *immunological tolerance*, *pattern recognition*, and so on viewed from the engineering standpoint.

Recent studies on immunology have clarified that the immune system does not just detect and eliminate non-self materials called *antigen* such as virus, cancer cells and so on, rather plays important roles to maintain its own system against dynamically changing environments through the interaction among *lymphocytes* and/or *antibodies*. Therefore, the immune system would be expected to provide a new methodology suitable for dynamic problems dealing with unknown/hostile environments rather than static problems.

Based on the above facts, we have been trying to engineer methods inspired by the biological immune system and the application to robotics [5][6][7][8]. We expect that there would be an interesting AI technique suitable for dynamically changing environments by imitating the immune system in living organisms.

In this chapter, we particularly pay close attention to the regulation mechanisms in the immune system, and propose a new decentralized consensus-making system based on the immune network architecture. We then apply our proposed method to behavior arbitration for an autonomous mobile robot, particularly to the *garbage-collecting problem* that takes into account of the concept of *self-sufficiency*. In order to verify our method, we perform some simulations and experiments. In addition, we try to incorporate adaptation mechanisms into the proposed artificial immune network based on *adjustment* and *innovation* mechanisms to autonomously construct appropriate immune networks.

2   Biological Immune System

2.1 Overview

The basic components of the biological immune system are *macrophages, antibodies* and *lymphocytes* that are mainly classified into two types, that is, *B-lymphocytes* and *T-lymphocytes*.

B-lymphocytes are the cells stemming from the *bone marrow*. Roughly 10^7 distinct types of B-lymphocytes are contained in a human body, each of which has distinct molecular structure and produces "Y" shaped antibodies from its surfaces. The antibody recognizes specific antigens, which are the foreign substances that invade living creature, such as virus, cancer cells and so on. This reaction is often likened to a *key and keyhole relationship* (see Figure 1). To cope with continuously changing environment, living systems possess enormous repertoire of antibodies in advance.

On the other hand, T-lymphocytes are the cells maturing in thymus, and they generally perform to kill infected cells and regulate the production of antibodies from B-lymphocytes as outside circuits of B-lymphocyte network (*idiotypic network*) discussed later.

For the sake of convenience in the following explanation, we introduce several terms from immunology. The key portion on the antigen recognized by the antibody is called an *epitope* (antigen determinant), and the keyhole portion on the corresponding antibody that recognizes the antigen determinant is called a *paratope*. Recent studies in immunology have clarified that each type of antibody also has its specific antigen determinant called an *idiotope* (see Figure 1).

2.2 Jerne's Idiotypic Network Hypothesis

Based on this fact, Jerne proposed a remarkable hypothesis called the *idiotypic network (immune network) hypothesis* [9][10][11][12][13]. This network hypothesis is the concept that antibodies/lymphocytes are not just isolated, namely they are communicating to each other among different species of antibodies/lymphocytes. This idea of Jerne's is schematically shown in Figure 1.

In the figure, the idiotope **Id1** of antibody 1 (**Ab1**) stimulates the B-lymphocyte 2, which attaches the antibody 2 (**Ab2**) to its surface, through the paratope **P2**. Viewed from the standpoint of **Ab2**, the idiotope **Id1** of **Ab1** works simultaneously as an antigen. As a result, Ab2 suppresses the B-lymphocytes 1 with Ab1.

On the other hand, antibody 3 (**Ab3**) stimulates **Ab1** since the idiotope **Id3** of **Ab3** works as an antigen in view of **Ab1**. In this way, the stimulation and suppression chains among antibodies form a large-scaled network and works as a self and non-self recognizer. We expect that this regulation mechanism provide a new parallel distributed processing mechanism.

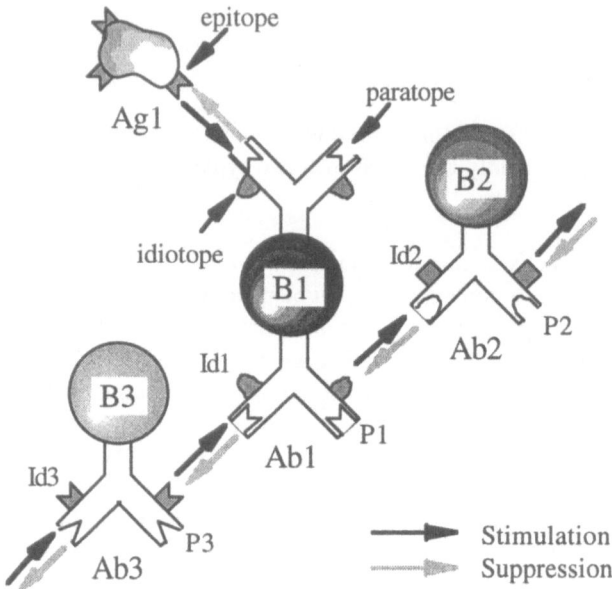

Figure 1. Jerne's idiotypic network hypothesis.

2.3 Metadynamics

In the biological immune system, the structure of the network is not fixed, rather variable continuously. It flexibly self-organizes according to dynamic changes of environment. This remarkable function, called the *metadynamics function* [14][15] [16], is mainly realized by incorporating newly generated cells/antibodies and/or removing useless ones. Figure 2 schematically illustrates the metadynamics function.

The new cells are generated by both gene recombination in bone marrow and mutation in the proliferation process of activated cells (the mutant is called *quasi-species*). Although many cells are newly generated every day, most of them have no effect on the existing network and soon die away without any stimulation. Due to such enormous loss, the metadynamics function works to maintain appropriate repertoire of cells so that the system could cope with environmental changes. The metadynamics function would be expected to provide feasible ideas to engineering field as an emergent mechanism.

Furthermore, new types of T-cell, which are also generated by gene recombination, undergo the selection in the thymus before they are incorporated into the body. In the selection mechanism, over 95% of them would be eliminated (*apoptosis*). The eliminated T-cells would strongly respond to self or not respond to self at all. In other word, the selection mechanism accelerates the system to incorporate new types effectively.

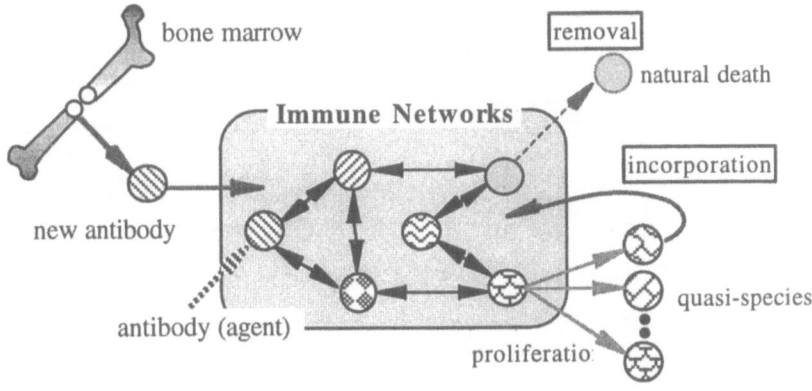

Figure 2. Metadynamics function.

3 Proposed Consensus-Making Network Based on the Immune System

3.1 Action Selection Mechanism and the Immune Network

As described earlier, in the behavior-based AI, how to construct a mechanism that realizes appropriate arbitration among the prepared competence modules must be solved. We approach to this problem from the immunological standpoint, more concretely with use of immune network architecture. Figure 3 schematically shows the action selection system for an autonomous mobile robot and the immune network architecture.

From the figure, we should notice that there are some similarities between these systems. That is, current situations, (e.g. distance and direction to the obstacle, etc.) detected by installed sensors, work as multiple antigens, and a prepared competence module (i.e. simple behavior) is regarded as an antibody (or B-lymphocyte), while the interaction between modules is replaced by stimulation and suppression between antibodies. The basic concept of our method is that the immune system equipped with the autonomous mobile robot selects a competence module (antibody) suitable for the detected current situation (antigens) in a bottom-up manner.

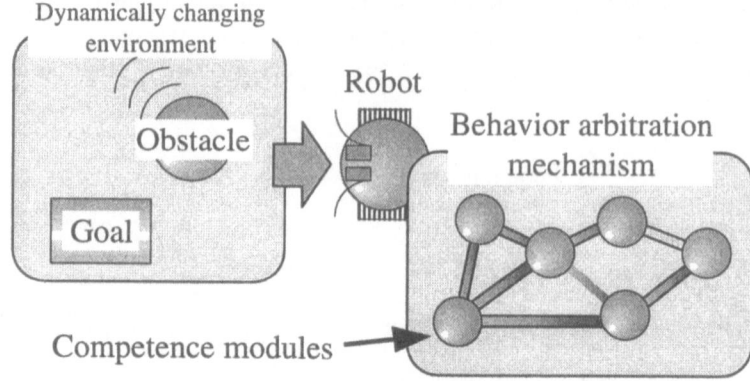

(a) An autonomous mobile robot with an action selection mechanism.

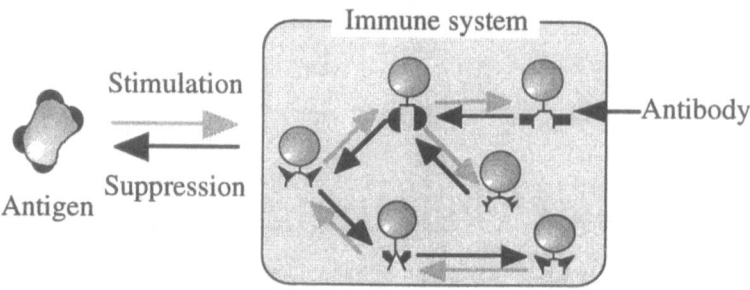

(b) Immune network architecture.

Figure 3. Basic concept of our proposed method.

3.2 The Problem

For the ease of the following explanation, we firstly describe the problem used to confirm the ability of an autonomous mobile robot with our proposed immune network-based action selection mechanism (for convenience, we dub the robot *immunoid*). To make immunoid really autonomous, it must not only accomplish the given task, but also be self-sufficient [17][18].

Inspired by their works, we adopt the following garbage collecting problem that takes into account of the concept of self-sufficiency. Figure 4 shows the environment. As can be seen in the figure, this environment, surrounded by walls, has a lot of garbage to be collected. And there exist garbage cans and a battery charger in the home base. The task of immunoid is to collect the garbage into the garbage can without running out of its internal energy (i.e. battery level). Note that immunoid consumes some energy as it moves around the environment. This is similar to the *metabolism* in the biological system.

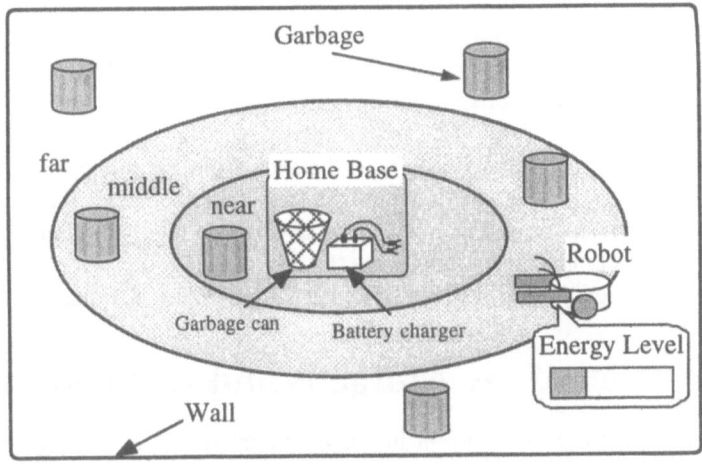

Figure 4. Environment.

In this study, we assume that prespecified quantity of initial energy is given to immunoid, and the simulated internal sensor installed in immunoid can detect the current energy level. For quantitative evaluation, we also use the following assumptions:

- Immunoid consumes energy E_m with every step.
- Immunoid loses additional energy E'_m when it carries garbage.
- If immunoid collides with garbage or a wall, it loses some energy E_c.
- If immunoid reaches the home base, it instantaneously gains full energy.
- If the energy level of immunoid is high, *go to home base* behavior might not emerge to avoid over-charging.

Based on the above assumptions, we calculate current energy level as:

$$E(t) = E(t-1) - E_m - k_1 E'_m - k_2 E_c \qquad (1)$$

$$k_1 = \begin{cases} 1 : garbage\ is\ in\ the\ hand \\ 0 : otherwise \end{cases}$$

$$k_2 = \begin{cases} 1 : collision\ with\ wall\ /\ garbage \\ 0 : otherwise \end{cases}$$

where $E(t)$ denotes the energy level at time t.

For ease of understanding, we explain why this problem is suitable for the behavior arbitration problem in detail using the following situations. Assume that immunoid is in the far distance from the home base, and its energy level is low. In this situation, if immunoid carries the garbage, it will run out of its energy due to the term E'_m in equation (1). Therefore, immunoid should select the *go to home base* behavior to fulfill its energy. In other word, the priority of the *go to home base* behavior should be higher than that of the *garbage-collecting* behavior.

On the other hand, if immunoid is in the near distance from the home base. In this situation, unlike the above situation, it would be preferable to select the *garbage-collecting* behavior. From these examples, it is understood that immunoid should select an appropriate competence module by flexibly varying the priorities of the prepared competence modules according to the internal/external situations.

3.3 Description of the Antigens and Antibodies

As described earlier, the detected current internal/external situation and the prepared simple behavior work as an antigen and an antibody, respectively. In this study, each antigen informs the existence of garbage (direction), wall (direction) and home base (direction and distance), and also current internal energy level. For simplicity, we categorize direction and/or distance of the detected objects and the detected internal energy level as:

- Direction   ->   front, right, left
- distance   ->   far, middle, near
- energy level   ->   high, low

Next, we explain how we describe an antibody in detail. To make immunoid select a suitable antibody against the current antigen, we must look carefully into the definition of the antibodies. Moreover, we should recall that our immunological arbitration mechanism selects an antibody in a bottom-up manner through interacting among antibodies.

To realize the above requirements, we defined the description of antibodies as follows. As mentioned in the previous section, the identity of each antibody is generally determined by the structure (e.g. molecular shape) of its paratope and idiotope. Figure 5 depicts our proposed definition of antibodies. As depicted in the figure, we assign a pair of precondition and behavior to the paratope, and the ID-number of the stimulating antibody and the degree of stimuli to the idiotope, respectively. The structure of the precondition is the same as the antigen described above.

We prepare the following behaviors for immunoid:
- move forward, turn right, turn left
- explore
- catch garbage
- search for home base.

For an antibody selection, an index that quantitatively represents suitableness under the current situation is necessary. We assign one state variable called *concentration* to each antibody (discussed later in section 3.5).

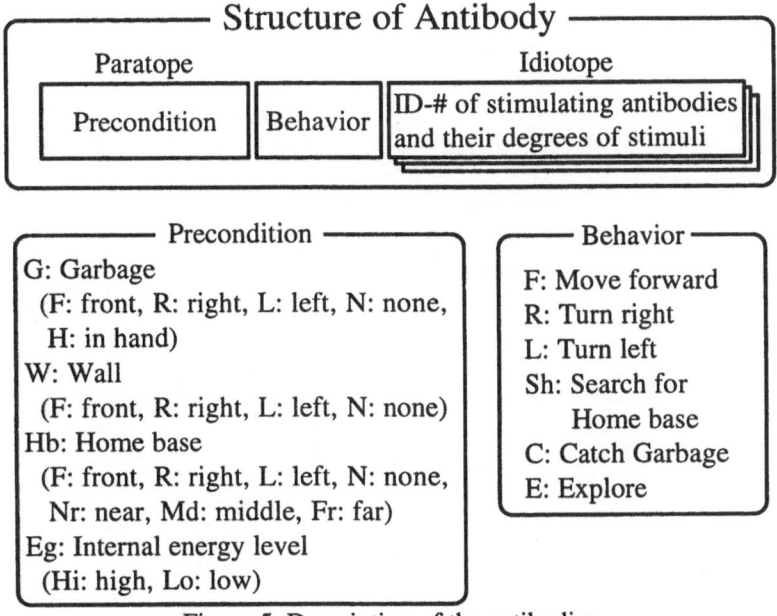

Figure 5. Description of the antibodies.

3.4 Interaction among Antibodies

Next, we explain the interaction among antibodies, that is, the basic principle of our immunological consensus-making networks in detail. For the ease of understanding, we assume that immunoid is placed in the situation shown in Figure 6(a) as an example. In this situation, three antigens possibly invade immunoid's interior.

Suppose that the listed four antibodies are prepared in advance that respond to these antigens. For example, **antibody 1** means that if immunoid detects the home base in the right direction, this antibody can be activated and would cause *turn right* behavior. However, if the current energy level is high, this antibody would give way to other antibodies represented by its idiotope (in this case, **antibody 4**) to prevent over-charging.

Now assume that immunoid has enough energy, in this case **antibodies 1, 2** and **4** are simultaneously stimulated by the antigens. As a result, the concentration of these antibodies increases. However, due to the interactions indicated by the arrows among the antibodies through their paratopes and idiotopes, the concentration of each antibody varies. Finally, **antibody 2** will have the highest concentration, and then is allowed to be selected. This means that immunoid catches the garbage (Figure 6(b)).

In the case where immunoid has not enough energy, **antibody 1** tends to be selected in the same way. This means that immunoid ignores the garbage and tries to recharge its energy. As observed in this example, the interactions among the antibodies work as a priority adjustment mechanism (Figure 6(c)).

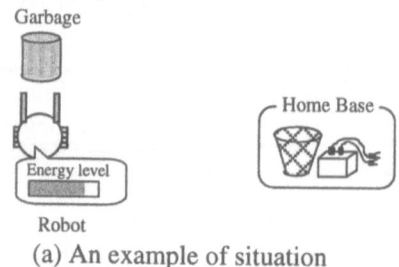

(a) An example of situation

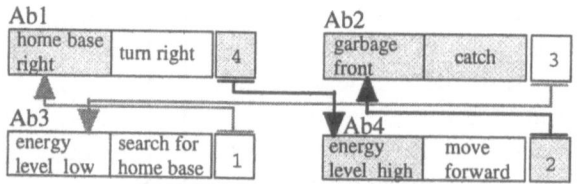

(b) In the case of high energy level.

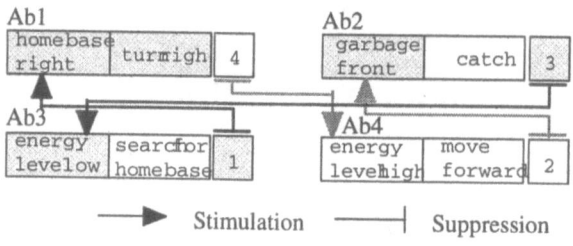

(c) In the case of low energy level.

Figure 6. An example of consensus-making network by interacting among antibodies.

3.5 Dynamics

The concentration of i-th antibody, which is denoted by a_i, is calculated as follows:

$$\frac{dA_i(t)}{dt} = \left\{ \sum_{j=1}^{N} m_{ji} a_j(t) - \sum_{k=1}^{M} m_{ik} a_k(t) - m_i - k_i \right\} a_i(t) \quad (2)$$

$$a_i(t+1) = \frac{1}{1 + \exp(0.5 - A_i(t))}, \quad (3)$$

where, in equation (2), N and M respectively denote the number of antibodies that stimulate and suppress antibody i. m_{ji} and m_i denote affinities between antibody j and antibody i (i.e. the degree of disallowance), and antibody i and the detected antigen, respectively. The first and second terms of the right hand side denote the stimulation and suppression from other antibodies, respectively. The third term represents the stimulation from the antigen, and the forth term the dissipation factor (i.e. natural death). Equation (3) is a squashing function used to ensure the stability of the concentration. In this study, selection of antibodies is simply carried out on a *roulette-wheel manner* basis according to the magnitude of concentrations of the antibodies. Note that only one antibody is allowed to activate and act its corresponding behavior to the world.

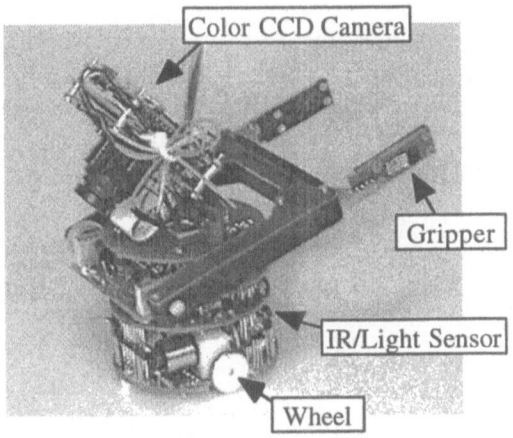

(a) Immunoid

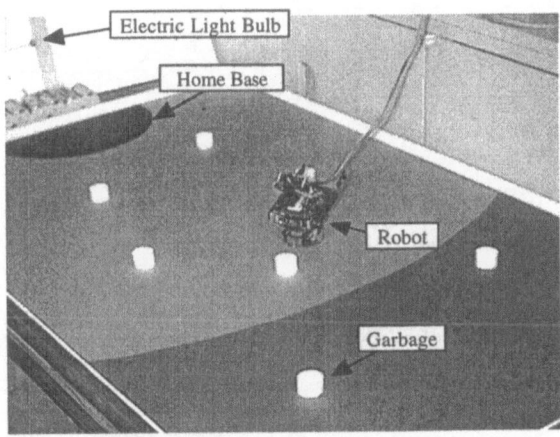

(b) Experimental environment

Figure 7. Experimental system.

3.6 Experimental Results

To verify the feasibility of our proposed method, we carried out some simulations and experiments. In the experiments, we used the *Khepera* robot, which is widely used for experiments, as immunoid. The robot has a gripper to catch the garbage and it is equipped with 8 infrared proximity sensors, eight photo sensors, and one color CCD camera. Each infrared sensor detects garbage or a wall of its corresponding direction. The photo sensors recognize the direction of the electric-light bulb (i.e. the home base). The CCD camera detects the color (red (far), green (middle), and blue (near)) at the current position and this in turn tells immunoid the current distance to the home base (see Figure 7).

As a rudimentary stage of investigation, we prepared 24 antibodies of which the paratope and the idiotope are described a priori (Figure 8). In the figure, the antibodies are categorized according to their roles. Figure 9 shows the structure of the immune networks by using the prepared antibodies. In the figure, the degrees of stimuli (affinities) in each idiotope are omitted for lack of space.

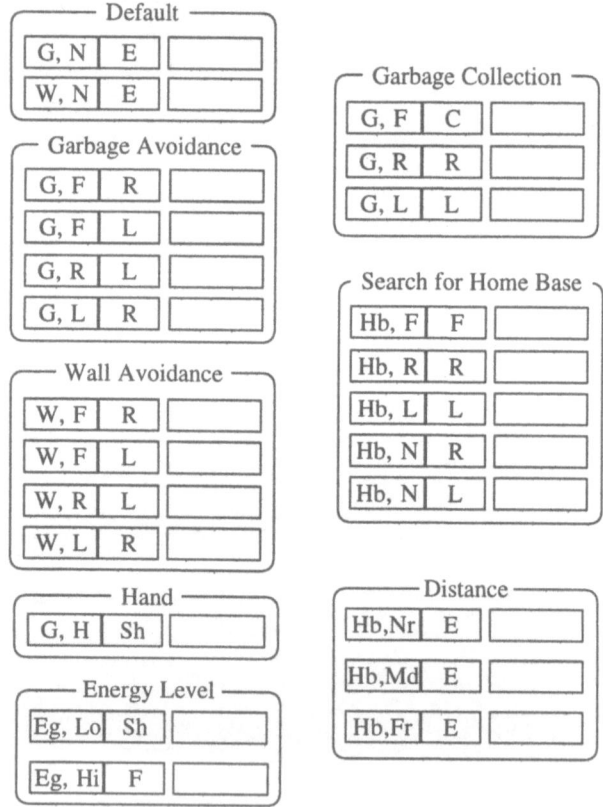

Figure 8. Prepared antibodies.

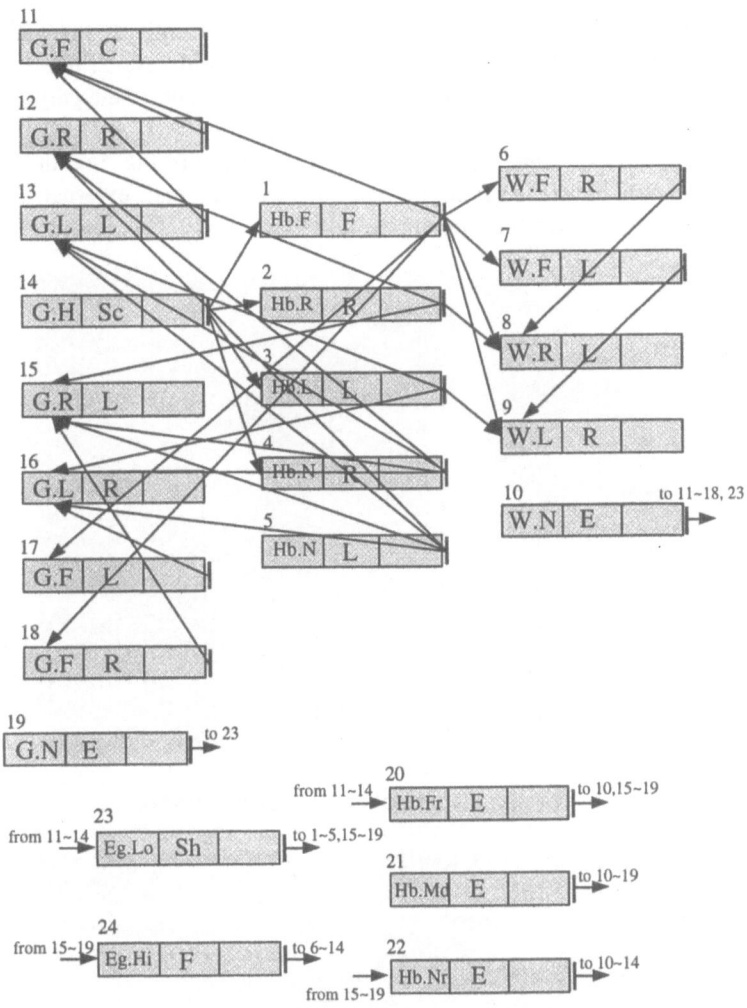

Figure 9. Structure of the immune network.

At the beginning, we equipped immunoid with the maximum energy level (i.e. 1000). We set $E_m=1$, $E'_m=3$, and $E_c=5$. Typical results obtained in the experiments are as follows: while the energy level is enough, immunoid tries to collect garbage into the home base. As the remaining energy runs out, immunoid tends to select an antibody concerned with *go to home base* and/or *search for home base* behaviors. After successful reaching the home base, immunoid starts to explore again. Such a regular behavior could be frequently observed in the experiments.

In order to evaluate the ability of our proposed arbitration mechanism, we furthermore carried out simple experiments by varying the initial energy level. Figure 10(a) and (b) are the resultant trajectories of immunoid in the case where the initial energy level is set to 1000 (maximum) and 300, respectively.

In case 1, due to the enough energy level, immunoid collects the garbage B and successfully reach the home base. On the other hand, in case 2, due to the critical energy level, immunoid ignores the garbage B and then collects the garbage A.

In spite of simple experiments, we can see from the figure that immunoid selects an appropriate antibodies according to both the internal and external situations by flexibly changing the priorities among the antibodies.

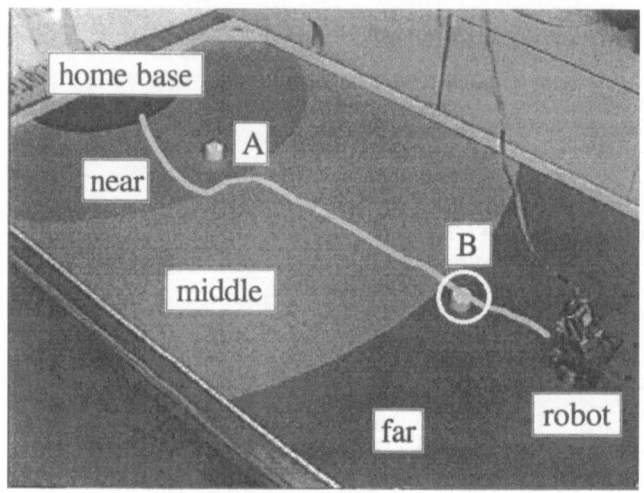

(a) Case 1 (initial energy level = 1000).

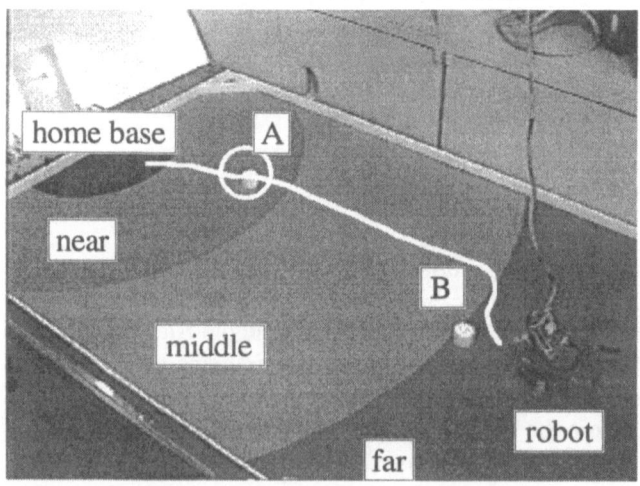

(b) Case 2 (initial energy level = 300).

Figure 10. Resultant trajectories.

4 Adaptation Mechanisms

4.1 Classification of Adaptation Mechanisms

In the previous section, we manually prepared antibodies and designed the immune network architecture. However, such a design approach would collapse as the number of the antibodies increased. Therefore, for more usefulness, the introduction of some adaptation mechanisms is highly indispensable. The adaptation mechanism is usually classified into two types: *adjustment* and *innovation* [19][20]. The adjustment can be considered as the adaptation by changing parameters in systems, e.g. modification of synaptic weights in neural networks, while the innovation as the adaptation by selection mechanisms. In the followings, we propose two types of adaptation mechanisms based on the adjustment and the innovation mechanisms.

4.2 Adjustment Mechanism

For an appropriate consensus-making, it is necessary to appropriately determine the ID-number of the stimulating antibody and its degree of stimuli m_{ij}, which are described in each idiotope. To realize this aim, we propose the on-line adjustment mechanism that initially starts from the situation where the idiotopes of the prepared antibodies are undefined, and then obtains the idiotopes using reinforcement signals.

For the following explanation, we assume that antigen 1 and 2 invade immunoid's interior (see Figure 11). In this example, each antigen simultaneously stimulates antibody 1 (**Ab1**) and 2 (**Ab2**). Consequently, the concentration of each antibody increases. However, since the priority between **Ab1** and **Ab2** is unknown (because idiotopes are initially undefined, there are no stimulation / suppression chain), in this case either of them can be selected randomly.

Now, assuming that immunoid randomly selects **Ab1** and then receives a positive reinforcement signal as a reward. To make immunoid tend to select **Ab1** under the same or similar antigens (situation), we record the ID-number of **Ab1** (i.e. 1) in the idiotope of **Ab2** and increase a degree of stimuli m_{21}. In this study, we simply modify the degree of stimuli as:

$$m_{12} = \frac{T_p^{Ab1} + T_r^{Ab2}}{T_{Ab2}^{Ab1}} \quad (4)$$

$$m_{21} = \frac{T_r^{Ab1} + T_p^{Ab2}}{T_{Ab2}^{Ab1}}, \quad (5)$$

where T_p^{Ab1} and T_r^{Ab1} represent the number of times of receiving penalty and reward signals when **Ab1** is selected. T_{Ab2}^{Ab1} denotes the number of times when both **Ab1** and **Ab2** are activated by their specific antigens. We should notice that this procedure works to raise the relative priority of **Ab1** over **Ab2**. In the case where immunoid receives a penalty signal, we record the ID-number of **Ab2** (i.e. 2) in the idiotope of **Ab1** and modify m_{12} in the same way. This works to decrease the relative priority of **Ab1** over **Ab2**.

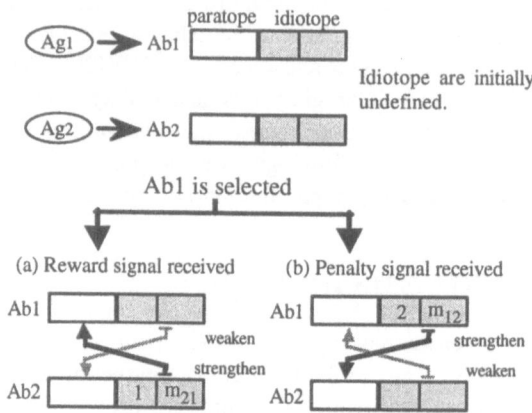

Figure 11. Proposed adjustment mechanism.

To confirm the validity of this adjustment mechanism, we carried out some simulations. In the simulations, the following reward and penalty signals are used:

Reward
- Immunoid recharges with low energy level.
- Immunoid catches garbage with high energy level.

Penalty

- Immunoid catches garbage with low energy level.
- Immunoid collides with garbage or a wall.

Figure 12 denotes the transitions of the resultant life time and collection ratio. From these results, it is understood that both are improved gradually as iterated. Figure 13 illustrates an example of the obtained immune networks through our proposed learning process. In this figure, for easy understanding, only the connections with high affinities ($m_{ij}>0.6$) are shown. From this figure, this network makes immunoid to search for the charging station when the current energy level is low. And it is also comprehended from this network that if the energy level is high, immunoid tends to select a garbage-collecting behavior. In spite of no explicit reinforcement signals concerned with distance to the home base, immunoid tends to search charging station if the distance to the charging station is far. While distance is near, it tends to collect garbage. We are currently implementing this mechanism into the real experimental system.

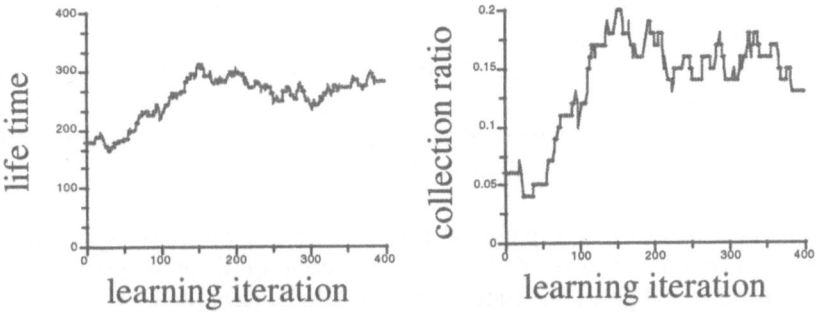

Figure 12. Transitions of life time and collection ratio.

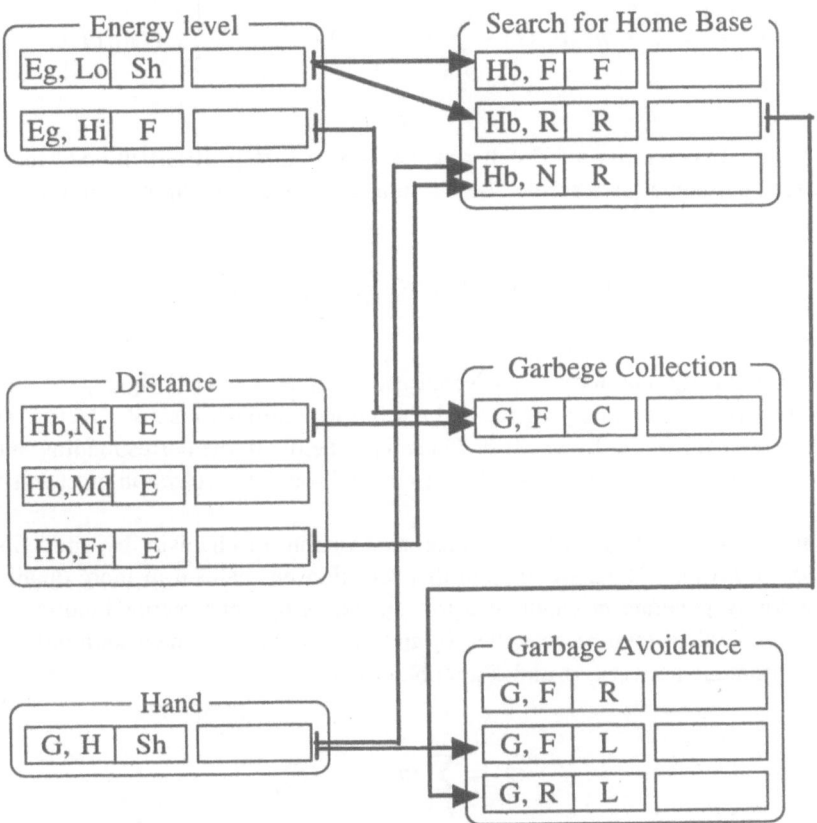

Figure 13. An example of the obtained immune network.

4.3 Innovation Mechanism

In the above adjustment mechanism, we should notice that we must still describe the paratope of each antibody in a top-down manner. One obvious candidate to avoid such difficulties is to incorporate an innovation mechanism. As described in section 2.3, in the biological immune system, the metadynamics function can be instantiated as an innovation mechanism. Therefore, we propose the following innovation mechanism inspired by the biological immune system.

Figure 14 schematically depicts the proposed innovation mechanism. Initially, the immune network consists of N antibodies, each of them is generated by gene recombination and given one state variable named concentration of B-cell. In order to relate this variable to the action selection process, we modify the equation (2) as follows:

$$\frac{dA_i(t)}{dt} = \left\{ \sum_{j=1}^{N} m_{ji} a_j(t) - \sum_{k=1}^{M} m_{ik} a_k(t) - m_i + k_i \right\} a_i(t) b_i(T) \quad (6)$$

$b_i(T)$ is the concentration of B-cell i in the T-th time step. If an antibody receives a reinforcement signal as a result of its action, the corresponding concentration of B-cell is varied as:

$$b_i(T+1) = b_i(T) + r_i \Delta b - K \quad (7)$$

where $r_i = 1$ if the antibody i is selected and receives a reward signal, $r_i = -1$ if the antibody i receives a penalty signal, and $r_i = 0$ if the antibody i is not selected. K is the dissipation factor of the B-cell. If $b_i(T)$ becomes below 0, the corresponding antibody is removed, then a new antibody is incorporated through the selection mechanism.

For quick improvement, we introduce a selection mechanism by modeling the function in *thymus*. Next, we explain this selection mechanism in more detail. First, we randomly generate m candidates for antibodies by gene recombination process. Then we calculate their sensitivities σ_i and δ_i between each new antibody and the existing immune network. σ_i and δ_i are obtained as:

$$\sigma_i = \sum_{j=1}^{N} m_{ji} a_j \quad (8)$$

$$\delta_i = \sum_{j=1}^{N} m_{ij} a_i \quad (9)$$

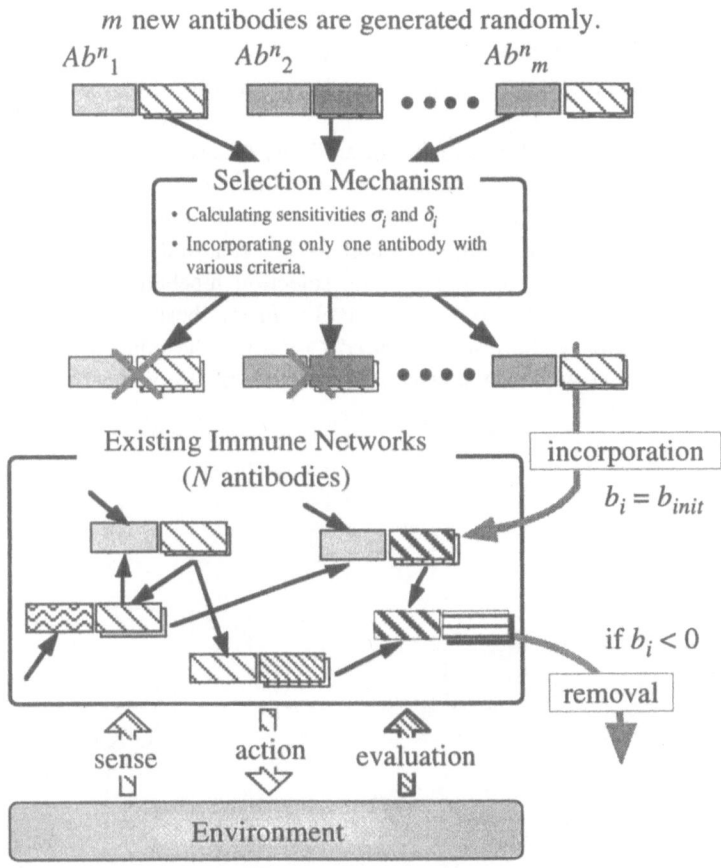

Figure 14. Proposed innovation mechanism.

As described earlier, each antibody has the interactions, i.e. stimulation and suppression. Sensitivity σ_i represents the sum of stimulation from the existing network, while sensitivity δ_i is the sum of suppression. Finally, only one antibody is allowed to be incorporated based on the predetermined criterion. In this study, we used $max\sigma_i$ and $max|\sigma_i - \delta_i|$ as criteria.

To confirm the ability of the proposed innovation mechanism, we applied to a simple example, i.e. obstacle avoiding problem. The simulated environment contains immunoid, multiple obstacles and one charging station. The aim of immunoid is to reach the charging station regularly in order to fulfill its energy level while at the same time avoiding collisions. In the simulations, the following reward and penalty signals are used:

Reward
- Immunoid approaches near the charging station with low energy level.
- Immunoid moves forward without collisions.

Penalty
- Immunoid collides with an obstacle or a wall.
- Immunoid does not move forward when there is no obstacle around it.

Additionally, we assume that the number of antibody N is set to 50, and the number of new antibody m to 20. Figure 15 denotes the transitions of the resultant lifetime, the number of move-forward actions and the number of collisions in three cases. In case (a), the selection mechanism is not used, namely one randomly generated antibody is incorporated, while in case (b) and (c), the selection mechanism is used with the criterion $max\sigma_t$ and $max|\sigma_t - \delta_i|$, respectively. From these results, the selection mechanisms (particularly the criterion $max\sigma_t$) improve the adaptation performance more rapidly than that without the selection mechanism. We are currently analyzing these results in detail.

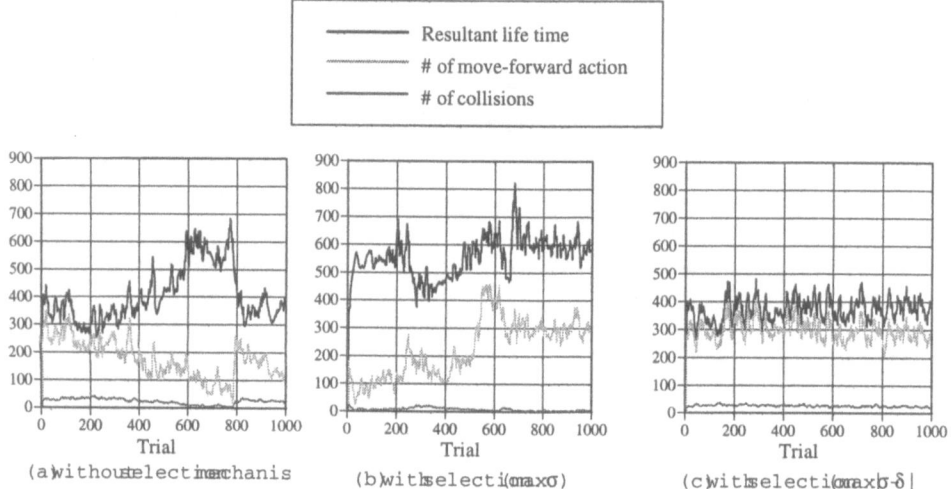

Figure 15: Simulation results under three different selection criteria.

5   Conclusions and Further Work

In this chapter, we proposed a new decentralized consensus-making mechanism based on the biological immune system and confirmed the validity of our proposed system by applying to behavior arbitration for an autonomous mobile robot. And we proposed two types of adaptation mechanism for an appropriate arbitration using reinforcement signals.

As artificial systems increase their complexities, the concept of self-maintenance and self-preservation become highly indispensable. For this, we believe that the immune system would provide fruitful ideas such as on-line maintenance mechanisms and robustness against hostile environments. This research is a first step toward the realization of artificial immune systems.

Acknowledgments

This research was supported in part by a Grant-in Aid for Scientific Research on Priority Areas from the Ministry of Education, Science, Sports and Culture, Japan (No.07243208, 08233208), and Mechatronics Technology Foundation.

References

[1]   R. Brooks (1986), "A Robust Layered Control System for a Mobile Robot", *IEEE Journal of R&A*, Vol.2, No.1, pp.14-23.

[2]   R. Brooks (1991), "Intelligence without reason", *Proc. of IJCAI-91*, pp.569-595.

[3]   P. Maes (1989), "The dynamic action selection", *Proc. of IJCAI-89*, pp.991-997.

[4]   P. Maes (1991), "Situated agent can have goals", *Designing Autonomous Agents*, MIT Press, pp.49-70.

[5]   A. Ishiguro, S. Ichikawa, and Y. Uchikawa (1994), "A Gait Acquisition of 6-Legged Walking Robot Using Immune Networks", *Journal of Robotics Society of Japan*, Vol.13, No.3, pp.125-128, 1995 (in Japanese), also in *Proc. of IROS'94*, Vol.2, pp.1034-1041.

[6]   A. Ishiguro, Y. Watanabe and Y. Uchikawa (1995), "An Immunological Approach to Dynamic Behavior Control for Autonomous Mobile Robots", in *Proc. of IROS'95*, Vol.1, pp.495-500.

[7]   A. Ishiguro, T. Kondo, Y. Watanabe and Y. Uchikawa (1995), "Dynamic Behavior Arbitration of Autonomous Mobile Robots Using Immune Networks", in *Proc. of ICEC'95*, Vol.2, pp. 722-727.

[8]   A. Ishiguro, T. Kondo, Y. Watanabe and Y. Uchikawa (1996), "Immunoid: An Immunological Approach to Decentralized Behavior Arbitration of Autonomous Mobile Robots", *Lecture Notes in Computer Science 1141*, Springer, pp. 666-675.

[9]   N.K. Jerne (1973), "The immune system", *Scientific American*, Vol.229, No.1, pp.52-60.

[10] N.K. Jerne (1985), "The generative grammar of the immune system", *EMBO Journal*, Vol.4, No.4.

[11] N.K. Jerne (1984), "Idiotypic networks and other preconceived ideas", *Immunological Rev.*, Vol.79, pp.5-24.

[12] H. Fujita and K. Aihara (1987), "A distributed surveillance and protection system in living organisms", *Trans. on IEE Japan*, Vol. 107-C, No.11, pp.1042-1048 (in Japanese).

[13] J.D. Farmer, N.H. Packard, and A.S. Perelson (1986), "The immune system, adaptation, and machine learning", Physica 22D, pp.187-204.

[14] F.J. Valera, A. Coutinho, B. Dupire, and N.N. Vaz. (1988), "Cognitive Networks: Immune, Neural, and Otherwise", *Theoretical Immunology*, Vol.2, pp.359-375.

[15] J. Stewart (1993), "The Immune System: Emergent Self-Assertion in an Autonomous Network", *Proceedings of ECAL-93*, pp.1012-1018.

[16] H. Bersini and F.J. Valera (1994), "The Immune Learning Mechanisms: Reinforcement, Recruitment and their Applications", *Computing with Biological Metaphors*, Ed. R. Paton, Chapman & Hall, pp.166-192.

[17] R. Pfeifer (1995), "The Fungus Eater Approach to Emotion -A View from Artificial Intelligence", *Technical Report, AI Lab, No. IFIAI95.04*, Computer Science Department, University of Zurich.

[18] D. Lambrinos and C. Scheier (1995), "Extended Braitenberg Architecture", *Technical Report, AI Lab, No. IFIAI95.10*, Computer Science Department, University of Zurich.

[19] B. Manderick (1994), "The importance of selectionist systems for cognition", *Computing with Biological Metaphors*, Ed. R.Paton, Chapman & Hall.

[20] J.D. Farmer, S.A. Kauffman, N.H. Packard, and A.S. Perelson (1986), "Adaptive Dynamic Networks as Models for the Immune System and Autocatalytic Sets", *Technical Report LA-UR-86-3287*, Los Alamos National Laboratory, Los Alamos, NM.

Chapter 3

REINFORCEMENT LEARNING OF A SIX-LEGGED ROBOT TO WALK AND AVOID OBSTACLES

Anne Johannet and Isabelle Sarda
LGI2P,
EMA - EERIE, Nîmes
FRANCE

Walking machine high potential for off-road or hostile environment mobility necessits an adaptive and versatile control systeme in order to avoid the difficulties of complex and unpredictible behaviour modelling.

Study presented in this chapter is devoted to a neural network based system control which allows a small six legged robot to walk and avoid obstacle even when it is partially dammaged. A very simple structure taking inspiration from insect behaviour is described. Reinforcement methods are investigated and appear to lead to an efficient learning.

1 Introduction

Autonomous Mobile Robots (AMR) are designed to assist humans in hard, monotonous or dangerous tasks. Researchers as well as engineers are interested in their development since they constitute an ideal domain for the implementation of artificial intelligence or control theories.

Design of an AMR is still an intricate task. Generally two types of problems are studied: the first one deals with locomotion (stability, efficiency) and the second one deals with guidance in order to reach a goal and to avoid obstacles. The major difficulty encountered in such a work is the extreme variability of the environment in interaction with the robot. Obviously nobody tries to realise a robot able to evolve in all types of surroundings, but the diversity of each situation is sufficient to lead to a relative failure with the classical modelling methods [1].

Different approaches have been investigated, and several tools used (cellular automata, genetic algorithms, etc....) in the design of AMR. The present study focuses on the connectionist reinforcement approach which allows the formalism of the supervised neural network to be applied in a partially modelled configuration.

After a presentation of the problematics of mobile robotics, linked to the necessity and to the organisation of the autonomous decision ability in Section 2, the learning methods using reinforcement learning are described in Section 3. Section 4 focuses on the description of the hardware platform used for experiments: a hexapod robot. At the end Section 5 investigates three types of behaviours: gait learning, obstacle avoidance, and walking in a damaged state. Each task was performed by a specific neural network; simulations as well as implementation on a real robot showed the advantages of this method in terms of simplicity, generalisability, and robustness.

2   Problematics of the Synthesis of an Autonomous Robot

2.1 Autonomy and Adaptability

The term "autonomous mobile robot" (AMR) corresponds to a wide variety of devices designed to operate where or when human agents cannot [2].

Space exploration is an obvious example of these situations, but they can also be found in many other domains at a lower technological level, but of wider industrial interest, for instance the transport of raw materials or finite products along a production line.

For all these types of systems, common features can be pointed out. An AMR is usually composed of three parts:

• Mechanical structure, with motorization and power source,
• Control command system,
• Carrying capacity.

A mobile robot is autonomous when it achieves the task it was designed for without any external energy, and furthermore when it is endowed with decision capabilities. These capabilities are a necessity when the robot is faced with changing environmental conditions because they enable it to determine its actions in order to remain in operation.

It is obvious that autonomy and adaptability capabilities are closely dependant. As its internal state and its surroundings continuously change, an autonomous robot has to continuously determine its next action in order to find the most appropriate way to perform its tasks. Strategies have to be defined to reach optimal conditions. This kind of adaptive control can be seen as the equivalent of physiological adaptation, *i.e.* biological term for physiological adjustment by an animal to external variations. Optimum action choice can be made either by comparison with a model, or by the learned results of past actions.

In the first process, the model definition requires a precise knowledge of the robot's environment, possible variations and influences on the robot's internal states. This knowledge may be hard to establish, especially when the robot is designed for outdoor operations.

Nevertheless, if we consider the navigation problem with obstacle avoidance, for instance, the question is whether it is worthwhile to define a complete and precise model for the robot environment. Observations of human navigation capabilities clearly indicate that the definition of exact position coordinates is not needed for moving towards a goal. A human agent gains from his or her sensors only information about the relative positions of goal and obstacles, and only when this information is useful [3].

In consequence, the second way to address the problem, based on trial-and-error experiments, seems to be more suitable for navigation problems.

2.2 Behaviour Based Control of AMR

When designing an AMR control system, two approaches are possible: the "classical approach", which was the first to be used, and the "behavioural approach", which has appeared more recently and relies on ethological considerations.

Classical approach
Robot sensors provide an embedded computer system with various data. Sensor data are fused into an environment model, used by the computer system to calculate successive actions.

Behavioural approach
Corresponds to an emerging trend of works on distributed control.

• The mobile agent is now defined by a set of behaviours, for instance "go forwards" , "get some food" and so on.
• Communication between the various behaviours is the key problem, since only one or a few of them can be active at a time and be in control of the robot actuators. So the main problem is to determine the prevailing behaviour.
• In most of the earlier works, behaviours form a hierarchical structure, which appears to be the most appropriate and leads to the lowest learning time.
• Brooks' studies were a precursor in this domain [4]. They used the so-called subsumption based approach, with the decomposition of the control problem into task achieving layers. Distributed networks of simple behaviours, that run

asynchronously and concurrently, are connected to form the task achieving layers.

2.3 Distributed Control Without Supervisor

As mentioned above, the adaptive process is based on the selection of emergent behaviour. This choice can be pre-programmed, as in Brooks' works. In this case, higher level behaviours can control lower level ones. Hierarchy between behaviours is defined by a human supervisor without any modification opportunity during operations. This approach may not be effective when the robot is exposed to an unpredictable environment.

The selection may also result from distributed control. An example can be found in Maes'work: there is no central supervisor to determine the hierarchical level of the different behaviours. Instead, local interactions between them determine the predominant behaviour at a given time [5].

A central supervisor is also found to be unnecessary in Beer's work. It is devoted to a simulated insect controlled by an artificial nervous system. Behaviour selection appears to be a global property of the nervous system [6].

2.4 The Present Work

The illustration of behavioural learning is performed on a six leg robot learning firstly to walk, and secondly to avoid obstacles. The particular case of the legged robot has been extensively studied, for it appears to be more appropriate for uneven ground than wheeled robots, and is less damaging to the natural environment. The problem of locomotion of a legged robot is to coordinate legs in order to adapt to a rough ground. Among the possible approaches, the one inspired by biological systems is interesting, particularly, as already mentioned, because it had been shown that the legs can be coordinated without any central supervisor [6].Our realisation of a six leg robot aims to validate these studies using reinforcement algorithm: given the aim simply to go forwards without falling down and without an exact knowledge of how to move, the robot adopts a behaviour in reaction to external stimuli. Concerning the hierachisation of the behaviour, we adopted the method proposed by R.A. Brooks in implementing both behaviours on two separate networks, each one specialising in one task: the first one in locomotion, and the second one in obstacle avoidance. The methods, choices and experimentation are reported below.

3   Reinforcement Learning

3.1 The Principle

Reinforcement learning originated in the observation of animal learning: when, in a given state, an animal reacts to an input in a satisfactory way, the particular association linking this input to its output is reinforced and will be delivered more frequently or more rapidly in the future. For example, a binary version of reinforcement learning can be the case of the reward-penalty learning: if the agent (animal or robot) performed a good action (relating to the input) it receives a reward, in the opposite case the response is a penalty. Of course, this type of learning has been studied by experimental psychologists, but it is also the subject of a growing number of works in the field of Engineering, especially in Robotics. For roboticians, the major advantage of the reinforcement methods is that the learning does not require a perfect modelling, neither of the robot 's actions, nor of the environment with which it interacts. Another way to consider this type of learning is that it allows a behaviour to be learned, and not a very precisely defined target function as usually occurs in control theory.

Another point of interest to consider is the selection of the inputs. Generally, in control theory, the problem is to find the input signal which brings the output nearer to the target; but in the case of reinforcement learning the problem cannot be stated in the same way because the models of both the agent action and the environment are only partially known. The proposed solution is therefore to introduce a stochastic element into the inputs in order to cause the robot to move in various directions, and then to allow it to learn the desired behaviour over a large part of the state space. The introduction of such noise has another benefit: when the environment or the agent characteristics change, the agent can adapt itself to these modifications in exploring once again the state space. Such a behaviour is shown in Section 5.3.

Implementing reinforcement learning for controlling an agent requires the following minimal configuration (Figure 1):

• An agent which:
 - receives the inputs coming from the environment (position, speed, É), and possibly feedback on its own state and the reinforcement signal in order to adapt.
 - proposes its actions.
• Measurement or modelling of the interactions between the actions proposed by the agent and the environment:
 - its inputs are the agent's proposed actions,
 - its outputs are the effective actions performed by the agent. The proposed actions and effective actions are not necessarily the same because of the

imperfect modelling of both partners: agent and environment. For example, a robot can slide on the floor or sink into soft ground.
• A heuristic function, which is a function taking information from sensors, and which has to evaluate the effects of the actions in relation to the desired goal (or behaviour). This heuristic function is called the "critic". The critic delivers the reinforcement signal which informs the agent of the quality of its actions.

Starting from the basic scheme as described, several problems emerged and led to adaptations in the learning process: in the case of complicated behaviour, it is necessary that the reinforcement signal gives more informations concerning the quality of the actions than a simple binary response. Generally this problem is bypassed by evaluating the quality of the response not only for the response given at one time step but also on a greater temporal window. The learning process under consideration is then a sequential process and lead not only to the evaluation of the current state, but also to the prediction of future states. Such approaches can be called "sequential reinforcement" approaches.

Moreover, there are different ways to make the agent. The early works, for example presented by Barto [7],addressed cellular automata models; currently the connectionist approach is very much used, and can also include the approach proposed by Widrow [8] operating on adaptive threshold systems. Moreover, some links with the problem of optimal control have been established, for example by Bertsekas [9] in Dynamic Programming. The connectionist approach is therefore presented below and the family of reinforcement learning which can be deduced from this approach, on one hand the one step method and, more quickly, the sequential methods.

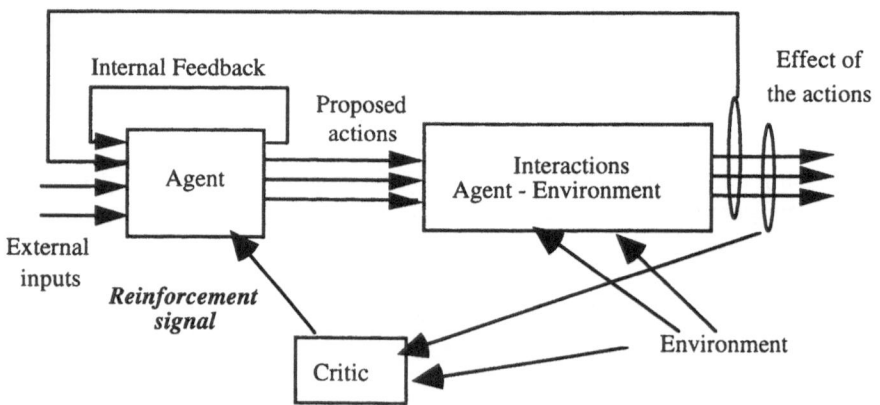

Figure 1. Learning by reinforcement.
The learning of the agent is performed using a reinforcement signal. This is computed by the "critic" from the environment configuration, and from the effect of the actions proposed by the agent. In order to inform the agent better, some feedback can be added on its own state.

3.1.1 The Neural Network: Definition and Learning

The most promising approach to making the agent is the connectionist or so called "neural networks" method because of its ability to identify any non linear process, as shown by Cybenko [10], and also to lead to intrinsically adaptive systems. This last property is a crucial one when we consider that the environment cannot be fully modelled and will oblige the agent to learn continuously in order to adapt to the environment's evolution.

An artificial neuron is a mathematical operator which generally computes two actions: first the linear weighted sum of its inputs, and second the non-linear evaluation of its output. Various models of neurons have been proposed depending on the evaluation function. One has :

$$s_i = f(\sum_j c_{ij}.e_j) \tag{1}$$

where s_i is the output of the neuron i, e_j is one of its inputs, and c_{ij} is the synaptic coefficient linking this input to the neuron under consideration; f(.) is generally a nonlinear function, for example it is possible to choose f(.) = tanh(.).

A neural network is a set of interconnected neurons. The connections between the neurons (defined by the set of coefficients c_{ij}) are computed during the learning procedure.

The neural network learning procedure can be viewed as the computation of the synaptic coefficients in order to minimise a "cost function". Different learning rules can be derived taking into account different cost functions and different minimising methods. Generally the learning is performed using a perfectly defined trajectory that the network's output has to follow. The learning is then called "supervised" learning because there is a reference that the network has to follow. The cost function can then be defined by the quadratic error between the effective outputs and the desired outputs of the network:

$$J(C,k) = \frac{1}{2}\sum_{\{o\}}\sum_{\{k\}}(s_o(k) - d_o(k))^2 \tag{2}$$

where $\{k\}$ is the set of input-output couples taken for k past values, $\{o\}$ represents the set of output neurons (neurons delivering the output of the network), and C is the set of synaptic coefficients.

In the case of stationary behaviour, the set of k values can include distant past events; but in the case of adaptive behaviour, as the plant's characteristics may change, only a few recent events are meaningful. Generally, minimizing the cost function is performed using a gradient descent. After the computation of the output s_o of the network, using a gradient method with a constant step μ, the modification to be applied to a particular coefficient connected to one output neurons is (at time t):

$$c_{oi}(t+1) = c_{oi}(t) - \mu . \frac{\partial J(C, t)}{\partial c_{oi}} \tag{3}$$

Nevertheless, other rules have been proposed, for example a gradient descent inspired by second order minimisation methods.

3.1.2 The Connectionist Reinforcement Learning

Connectionist reinforcement learning is the adaptation of the reinforcement principle to neural networks. Such adaptation can be done easily starting from the stochastic automata proposed by Barto [7]: the authors were inspired by two previous algorithms: first the *stochastic learning automaton algorithm,* where the automaton's behaviour is governed by a probabilistic evaluative feedback of the environment, and secondly from the *supervised learning pattern classification* algorithm which minimises a cost function via a gradient descend method. The link with the connectionist approach can be made in substituting the stochastic automaton by artificial neurons; therefore the agent is modelled by a neural network. The learning of the network is implemented with the cost function used in the *supervised learning pattern classification* algorithm.

3.2 The Family of Reinforcement Learning Rules

3.2.1 One Step Reinforcement Signal

The framework in which the connectionist reinforcement has been introduced consists of a neural agent and an environment connected in a feedback loop. The agent produces an action, and the environment responds to this action by an immediate evaluative signal which is usually interpreted as a success or a failure.

• *Associative learning rules*
Depending on whether the inputs of the agent are only the reinforcement signal or this signal and external inputs, the learning method is called respectively a "non associative" or "associative reinforcement" one step learning rule.

Inspired by Barto [11] a general presentation of the associative reward penalty algorithms (A_{rp}) can be presented in an interesting way. Let us consider the case of the associative learning rule, a very promising approach has been proposed by Gullapali [12]:

• First the stochastic aspect of the state space exploration is obtained by adding a noise to the weighted sum of the neurons, to obtain (N is the noise):

$$s_o = f(N + \sum_j c_{oj} e_j)$$

• Second: when the reinforcement signal shows that the movement produced by the agent was appropriate, the learning must incite the agent to adopt this behaviour, *i.e.* must increase the probability of the agent to redo the same action.

This last idea can be translated in terms of desired or supervised value. Let us consider for that the case of a binary evaluation, the reinforcement signal can be taken equal to 1 for a reward and to -1 for a penalty, as:

• If the output s_0 is appropriate (r=+1), then the desired output is s_0 or $r.s_0$ (redo the same movement),
• If the output s_0 is inappropriate (r=-1), the desired output is $-s_0$, or also $r.s_0$ (opposite movement).

Moreover, the influence of the noise must be taken into account to calculate different stochastic inputs applied to the agent but must not be used to perform learning itself. It is therefore necessary to cause, not exactly the output to reach the desired value, but the output value without the noise effect. To do that, we consider the mean value of the output operated on an infinite number of trials. Hopefully, if the noise is suitably chosen it is possible to compute the expectation of outputs without redoing an infinite number of trials. For example, considering binary neurons with a threshold as evaluation function, and defining the noise as follows:

$$s_0 = +1 \text{ with the probability: } P(+1) = \frac{1}{1 + \exp(-2\beta v_o)}.$$

$$s_0 = -1 \text{ with the probability: } P(-1) = \frac{1}{1 + \exp(+2\beta v_o)}.$$

where v_0 is the weighted sum of the inputs (e_j), computed by an output neuron, the estimation of the expectation value can be computed without statistics:

$$E(s_0 | c_{oj}, e_j) = (+1).P(+1) + (-1).P(-1) = th(\beta.v_j).$$

The analogy with classical supervised way of performing learning in neural networks can then be done in using:

• $E(s_0 | c_{oj}, e_j)$ for the computed output value of the network (s_0),
• $r.s_0$ for the desired output value of the network (d_0).

It is then possible to derive a family of learning rules as for pattern recognition problems. For example, the Hebb rule can be translated as:

At time t, the increment to apply to a coefficient is:

$$c_{oj}(t+1) = c_{oj}(t) + \mu.E(s_o|c_{oj}, e_j).e_j(t).$$

Another well-known rule called LMS (linear mean square) or Widrow-Hoff rule has been extended to the reinforcement context by Widrow himself [8] without the stochastic part described previously. Applying to the LMS the stochastic process, the expression is then:

At time t, the increment to apply to a coefficient is :

$$c_{oj}(t+1) = c_{oj}(t) - \mu.(E(s_o|c_{oj}, e_j) - r.s_o).e_j(t).$$

It is easy to note that, if the learning converges, as expected the increment tends to zero and then the value $E(s_o|c_{oj}, e_j)$ tends to the "desired value" $r.s_o$.

Although similar to this expression, the A_{rp} algorithm proposed by Barto still differs in one aspect: the gradient step μ is different for positive and negative reinforcements. Although this difference could appear to be small, its importance is crucial because generally at the beginning of learning, there are many more inappropriate actions than appropriate actions and, consequently the network is always learning not to redo the inappropriate actions and doesn't learn the appropriate actions. Consequently, and as shown in Section 5.1. the gradient step has to be chosen greater for a reward than for a penalty.

One convergence theorem had been proved by Barto [7] in the case of linearly independent input vectors and in the case of a two-action-automaton. Extension of this theorem to several automata (several neurons) is not trivial and has not, to our knowledge, yet been done.

3.2.2 Sequential Reinforcement Methods

Generally the actions performed by the agent have some consequences not only on the next state but also on all successive states. In this case it is logical not to cut down arbitrarily the evaluation of the behaviour to the next actions and to take into consideration several actions. Of course, because the evaluation of the actions is taken into account to compute the reinforcement signal, a predictor is necessary for both the proposed action of the agent and also for the action's effect. Although this way might appear inconsistent with the hypotheses of a partially known environment, the method can be successfully used when the process is complicated by taking into account few predicted terms with an attenuation over time. In the simplest case, the objective of learning is then to select the actions at time t which maximise the rewards over all time steps in the future:

$$R(t) = \sum_{k=0}^{\infty} \gamma^k . r(t+k).$$

where r(t) is the reinforcement signal at time t, and $0 < \gamma < 1$ is the coefficient allowing to decrease progressively the influence of long time predictions.

In the connexionist approach, the prediction of the future reinforcement signals is naturally done by a neural network; the parallel with the temporal difference method introduced by Sutton [13] is presented below.

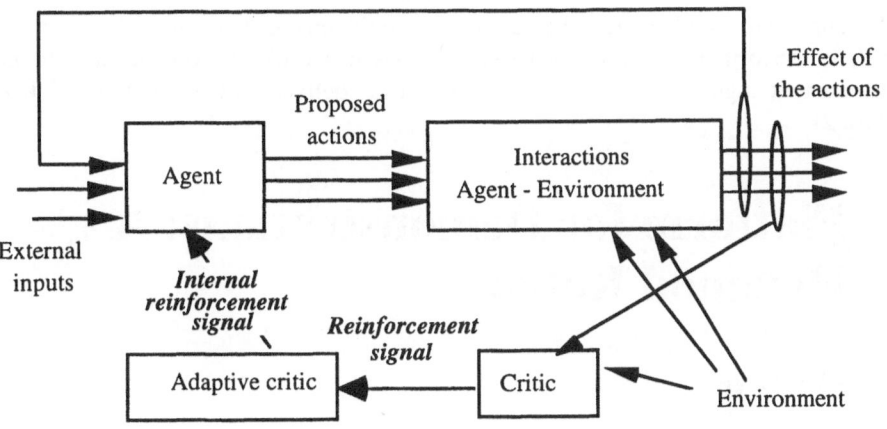

Figure 2. Reinforcement using adaptive critic.

Let us consider a supplementary neural network devoted to the prediction of the cumulated reinforcement R(t). This neural network is called the adaptive critic. R(t) can be written as follows:

$$R(t) = \sum_{k=0}^{\infty} \gamma^k . r(t+k),$$

and

$$R(t+1) = \sum_{k=0}^{\infty} \gamma^k . r(t+1+k).$$

Obtaining:

$$R(t) - \gamma R(t+1) = r(t).$$

In this expression, at time t, r(t) is known, R(t) is the unknown value that we want the adaptive critic to learn how to predict (output of the network), and R(t+1) is an estimation computed by the adaptive critic with the inputs at time t+1 (available after the computation at time t) using the synaptic coefficients available at time t.

A learning rule inspired from LMS can then be derived.

At time t, the increment to apply to a coefficient is :

$$c_{oj}(t+1) = c_{oj}(t) - \mu.(r(t) + \gamma R(t+1) - R(t)).e_j(t).$$

In this approach, the agent uses the cumulated predictions of (R(t)), the reinforcement signal, to perform its learning, whereas the classical critic delivers its immediate reinforcement signal r(t) and the adaptive critic delivers the prediction of R(t) (Figure 2).

4 Platform for Demonstrations: A Hexapod Robot

4.1 Design

4.1.1 Choice of the Platform

A hexapod robot was chosen to evaluate the properties of reinforcement learning due to two considerations: on the one hand the goal of the design of a hexapod robot is to build an autonomous vehicle ableto evolve in a complicated or partially unpredictable natural environment; on the other hand the control of the movement is not a trivial one due to the coordination between the six legs. The particular choice of six legs and not eight or four legs was made because it leads in a first stage to a simplification of the real device since walking can be implemented without taking into account the dynamical model of the robot. When walking, the gait is stable if the centre of gravity is always inside the basis of support. Restriction to a walk as opposed to a trot or a gallop), essentially limits the speed of the robot.

4.1.2 Input-Output Coding

As explained in Section 3.1, the *control system* is designed using a neural network. Basically the first idea for coding the inputs and outputs of the network was to encode the positions of the six legs. The electromechanical control of the legs implementing a walk with constant step length, there were four positions for each leg: up and behind, down and behind, up and ahead, down and ahead. Coding this problem leads to a neural network of $6*2 = 12$ inputs and 12 outputs of binary neurons.

Consequently the search for appropriate movements using a reinforcement learning is done in a 2^{12} dimensional space which seemed to be too large considering the type of problem. We therefore decreased this state space by choosing another way of coding the problem: the output of the neurons encode not the positions of the legs, but the movements: move forward or move backward. In this way, the state space dimension decreases to 2^6 positions which is more reasonable. Consequently, the time of convergence would be decreased, the quality of the behaviour would also be increased. When at least one leg on each side of the robot is put backward, the robot takes a step. The neural network computes parallel iterations, then all movements are synchronous.

The block entitled "Interactions Agent-Environment" in Figure 1 is, in the case of a real robot, *de facto* implemented by the robot itself. Nevertheless we includes a supplementary filter in the software in order to protect the electromechanical device: all the steps having the same length, if the output of a neuron is identical two times successively, the movement cannot be redone, it is inhibited.

4.2 Hardware Description

The electromechanical design was constrained according to dimensions, weight, architecture, cost and mechanical and electrical power. Sensors provide information to the robot about its legs positions, its advancement, and if it falls. The robot is 27 cm long and 15 cm wide (Photo in Figure 3). All the sensors, signal processing and motors control electronics are on board. Only the power source is external and comes to the robot via a cable. It is therefore autonomous not in terms of power but in terms of computation and decision.

Principal specifications of the robot are listed in Table 1.

Table 1 Specifications of the robot

Mechanical characteristics		Electronic characteristics	
Weight (mechanics)	2 kg	Power supply for logic and control	5V, 3A
Weight (electronics)	1 kg	Power supply for engine	±15V, 4A
Clearance shoulder	30ⱼ	Microprocessor	80188
Clearance elbow	45ⱼ	Ram	128 Ko
Power engine	3.8 W	Eprom	128 Ko
Reduction shoulder	33	Serial links	2
Reduction elbow	1090	Interrupt links	2
Inc. encoder nbr of stroke	16		
Switch detector	12+8		

The hardware architecture is centred around a bus connecting one microprocessor card, six control cards and one sensor card. Motor control is performed using a specialised digital circuit (National Semiconductor LM628).

The software is organised in two parts: one, resident, for libraries, and the other, downloaded (learning algorithm). It was developed in C language.

These high capabilities allow first the control of amplitude and speed of the movement, and secondly the management of exteroceptive sensors. They should give us the tool to investigate some problems such as the trajectory curves, crossing obstacles or changing gait.

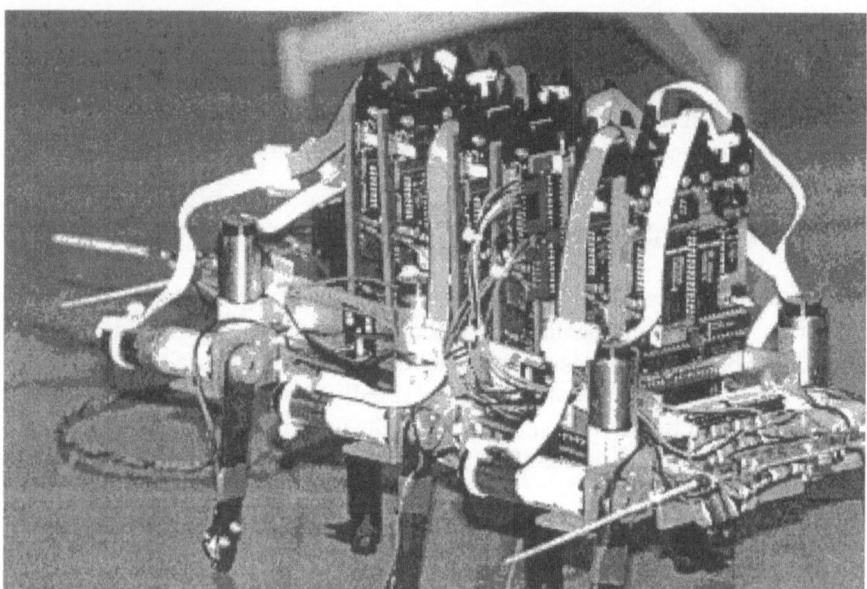

Figure 3. The hexapod robot.

4.3 Real World-Virtual World Interaction

Two studies were performed concurrently: the first one is the electromechanical conception of the robot and the second one is the realisation of the software, to simulate learning in a first step, and then for the control of the robot. In order to minimise the differences between the two aspects of virtual and real world we took into account some interactions :

• Initial position of the robot.
• Position after a fall.
• Static model of the robot.
• Forbidden movements: these movements, avoided by "muscular" filters had been
    distinguished in two categories: those which break the robot, implemented for

virtual and real robot, and those which are physically impossible to do (move the robot forward with only one leg...), implemented only on the virtual robot. These filters are included in the block converting the agent proposed actions into effective actions.

4.4 Sensor Definition

Over and above the sensors and detectors needed to ensure correct position control of legs and correct initialisation (leg-ground contact), supplementary sensors are needed to allow the robot to behave autonomously. For that, two types of sensors have were embedded in the robot.

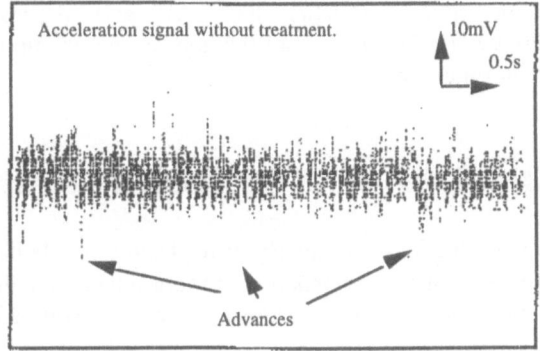

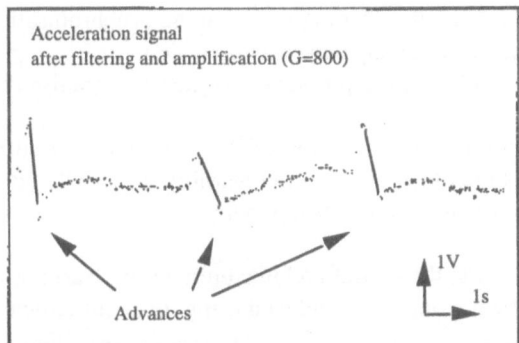

Figure 4. Signal processing for acceleration detection.
Band pass filter and high gain (800) allow us to extract a working signal.
Acceleration detection can then be made on the robot as a proprioceptive sensor.

Exteroceptive detectors:
- Fall detectors, which are simple micro-switches situated under the body of the robot.
- Obstacle detectors of obstacles which are inspired by cat's whiskers, implemented in the real device with long sticks operating on micro-switches.

Proprioceptive sensor:
• An acceleration sensor in order to know if the robot is advancing or not.

The easiest solution to have information of advancement was to use several beacons and to detect the movement by using, for example, ultra-sound signals. However we rejected this solution considering that it was incompatible with the goal of "autonomous behaviour in an unpredictable environment", as discussed previously. We therefore decided on internal detection of the movement, operated by a kind of proprioceptive sensor detecting acceleration. As one can imagine, the acceleration sensor was the most delicate to deal with because of firstly the very small values of the acceleration induced by the robot's movements, secondly the interferences induced by the walking movements themselves This problem was solved using a very sensitive acceleration sensor (Kistler 83038A), and processing the signal appropriately: analogue filtering and amplification. Some examples of signals thus obtained are shown in Figure 4.

5   Learning of Three Behaviours

Having on the one hand the hardware platform for testing real behaviour, and on the other hand, the software platform to experiment more intensively the convergence of learning and the influence of the parameters, we investigated three types of behaviour:

• First the six legged walk, in order to prove that the synchronisation between the legs could be learnt without any central supervision.
• Second obstacle avoidance, the problem was then to organise the both behaviours: walking and voidance.
• Third the adaptation to a modification of the robot itself, occurring with a failure. The goal was then to prove that the learning was really adaptive and can then lead to effective autonomous behaviours.

For the three behaviours, the neural architecture and the learning algorithm were the same, the different behaviours were induced using different critics.

5.1 Definition of the Neural Network

5.1.1 Architecture of the Network

The proposed global architecture includes two separated networks. Both networks are based on the same architecture, and use the same learning rule, only their critic functions are different. The basic network is composed of six neurons arranged in one layer as a perceptron. Each neuron is affected to one particular leg. Globally, the neuron's output code the movement whereas the inputs receive the position of the legs (Figure 5).

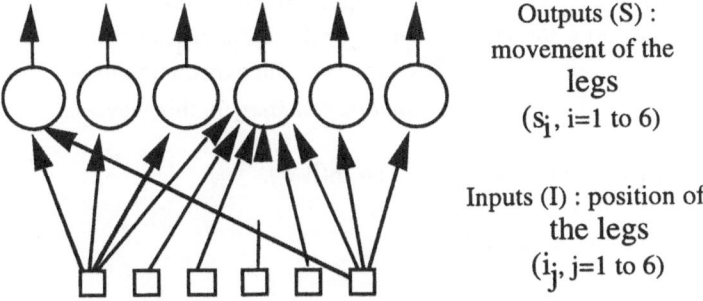

Outputs (S) :
movement of the
legs
$(s_i, i=1$ to 6$)$

Inputs (I) : position of
the legs
$(i_j, j=1$ to 6$)$

Figure 5. Neural network delivering the movement.
Neurons are shown as circles, their inputs are visualised as squares and the coefficients figure as links. Each neuron is connected to all inputs. The connections of one neuron are fully shown in black links, other connections are not totally visualised (some of them appear as grey links). The neural network contains only one layer of neurons.

The iterations of the network are parallel. Each iteration is decomposed as follows:

• For each neuron (i = 1 to 6), the weighted sum is computed:

$$v_i = \Sigma_{j=1,6} \, c_{ij} \cdot i_j,$$

where the parameters c_{ij} are the synaptic coefficients, i_j is the input delivering the position of the leg j, and v_i is usually termed the "potential", by analogy with the biological post-synaptic potential.

• For each neuron, its output (or its action) is computed. The output is evaluated by applying a noisy threshold to the potential, using:

$$s_i = sign(v_i + b_i),$$

where sign(.) is the function sign and b_i is a noisy bias. As described in section III.2.a., the noise is chosen in order to obtain the binary output (s_i) depending on the potential value :

$$s_i = +1 \text{ with the probability: } P(+1) = \frac{1}{1 + \exp(-2\beta v_o)}.$$

$$s_i = -1 \text{ with the probability: } P(-1) = \frac{1}{1 + \exp(+2\beta v_o)}.$$

β is a parameter which has to be fixed before learning.

5.1.2 Learning

The learning procedure of a neural network consists in the computation of the synaptic coefficients (c_{ij}, i and j = 1,6). As proposed by Touzet [14] for a similar task on a software platform, the chosen learning rule is the A_{rp} algorithm introduced in section 3.2.1. The coefficient's modification uses the expectation of the output s_i, this value can be computed (*and not estimated by statistics*) in this way:

$$E(s_i|c_{ij}, i_j) = (+1).P(+1)+(-1).P(-1) = th(\beta.v_i).$$

Then the learning rule can be written as:

If r = +1 (reward), $\Delta c_{ij} = \mu^+(r.s_i - E(s_i|c_{ij}, i_j))i_j$.
If r = -1 (penalty), $\Delta c_{ij} = \mu^-(r.s_i - E(s_i|c_{ij}, i_j))i_j$.

μ^+ and μ^- are the two parameters controlling the step of the increment.

At the beginning of learning, the coefficients are initialised so as to have equiprobability of movement for each neuron (Prob=0.5) in order to explore the space state. At each iteration the reinforcement signal (r) is taken into account and modifies the coefficients in order to increase the probability of "right" movements. The probability to obtaining a "right" movement is reported versus the potential value in Figure 6. It appears that the probability is effectively equal to 0.5 for a potential near to zero, and is high for a great potential which is generally obtained at the end of learning. So, when the procedure works well, at the end of learning, the behaviour is deterministic.

As β, μ+ and μ- have to be chosen before learning, their role will be described in a future section.

5.2 Gait Learning

The network architecture and the learning rule have already been described, the specificity of the procedure leading to the gait learning is specified in the critic function. The critic function is a heuristic one which converts the global goal (or behaviour) into easily measurable sub tasks. For example in the case where we want the robot to learn how to walk, the critic function can be :

• *Advance* which can be expressed as: *do not fall down, do not stay immobile*.

The evaluation of the both tasks: not to fall down and not to stay immobile, is performed through sensor's outputs, after the movement's completion. This evaluation is never performed based directly on the proposed actions (the output of the network), but on the effect of the actions.

Reward and penalty can be explained as follows: after one movement, if the robot fails (falls down or stays immobile), the previous movement was wrong, then the robot is encouraged to do the opposite movement. If the robot takes a step, it is encouraged to do the same movement.

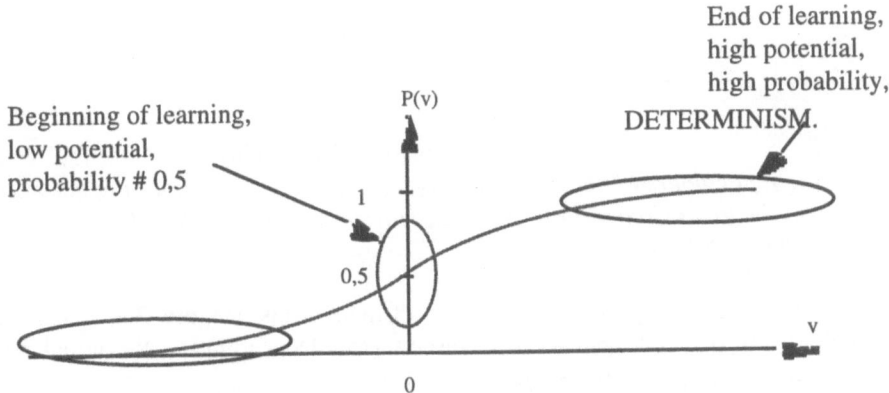

Figure 6. Influence of the noise on the learning

The probability to obtain an output equal to one is drawn versus the potential value (weighted sum). At the beginning of the learning, as the coefficients are randomly initialised to small values, the potential is approximately equal to 0; then the probability that the output of the neuron will be equal to 1 is approximately 0.5: movements are made randomly. On the other hand, at the end of learning, the potential has reached a great value (#0 or #1) and the probability tends consequently to be equal to 0, or 1 respectively. The process is then deterministic as long as the agent and the environment are stationary.

5.2.1 Tripod Walk

Applying the A_{rp} algorithm, and using the critic defined before, when the parameters are well adjusted, and if the robot is initialised in a stable position, the learning converges towards a solution, in less than 100 parallel iterations, with a probability of 0.7. At the beginning of procedure, the robot moves its legs randomly. The movements which lead to an advance are encouraged and the probability of their occurrence increases. The movements which lead to a fall are progressively eliminated.

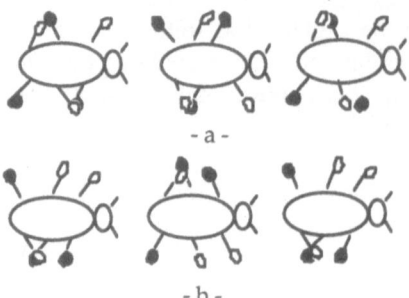

Figure 7. Walks found by learning.
The foot coloured in black represents a leg on the ground (pushing backward),
and the foot coloured in white, a leg moving forward. Both walks are periodic
with two states.

The most frequently found walk is a "tripod" walk (Figure 7.a). This walk is biologically plausible and used by insects [15]. Depending on the initial position, another walk (Figure 7.b) can occur.

5.2.2 Learning Convergence

The convergence towards a hexapod gait depends on three parameters: the slope of the probability (β), and the two learning parameters μ^+ and μ^-. The role of these parameters and of the initialisation is described below. Due to the permanent learning and the possibility to have an unprobable movement, the convergence cannot be stated rigorously; nevertheless we decided that the network has converged when twenty steps were perfectly performed without falling down or immobility. On the whole set of trials we performed, this criterion always appeared pertinent.

Influence of $\mu^{\pm}$

The relative magnitude between μ^+ and μ^- has to reflect the ratio between the number of states contributing positively (μ^+) and the number of states contributing negatively (μ^-) in the state space during the beginning of learning. In the case of gait learning, there are fewer "right" than "wrong" actions, therefore we have to choose $\mu^+ >> \mu^-$.

When the learning works well, the convergence of the coefficients evolve according to two behaviours: the first one corresponds to a movement leading to a penalty: this case is more frequent and induces the coefficients to evolve smoothly (small slope on the figures); the second case corresponds to a "good" movement: this case is rare and leads generally (depending on the term $(r.s_i - E(s_i|c_{ij}, i_j))$) to a great increment (large step on the figures : $\mu^+ >> \mu^-$). The convergence always follows a reward increment.

Depending on the initialisation and stochastic process, the learning search into the output state space is never identical: an example of two searches is shown (Figure 8).

If the ratio between parameters μ^+ and μ^- is not good the learning procedure may not converge (Figure 9).

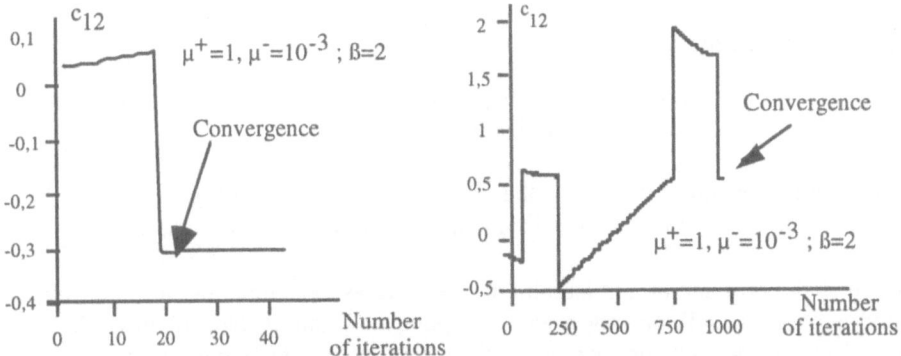

Figure 8. Convergence of a coefficient.
The number of iterations is the number of movements. This coefficient evolution corresponds to learning with a good choice of parameters: $\mu^+ >> \mu^-$ $(=10^3)$. The convergence occurs on the left after about 20 iterations. On the right, the learning stochastic evolution leads in this case to a slow convergence which is relatively rare.

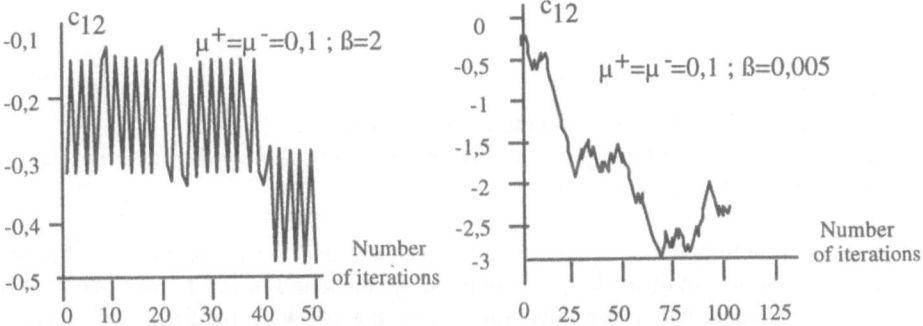

Figure 9. Evolution of a coefficient without convergence
On the left, an example of bad evolution of "learning" because of the inappropriate choice of parameters. The "learning" can't converge because the penalties are preponderant relatively to the rewards. On the right, the slope β is very low, so the noise is greater than the potential of the neuron: the behaviour of the learning procedure is essentially stochastic; there is no convergence.

5.2.3 Influence of the Noise

The higher the slope of the probability function, the lower the effect of the noise. The noise is crucial because it allows the exploration of the output state space in order to find an original solution. If the slope is too high, the network is deterministic and does not learn (cycles). If the slope is too shallow, the learning process does not converge (Figure 6). The magnitude of the slope is estimated in relation with the amplitude of the parameters μ^+ and u .

5.2.4 Conclusion

It appears in this study that learning how to walk can be performed using a reinforcement method in a very efficient way:

• If the parameters are well chosen, the convergence is always reached.
• In 70% of trials the convergence occurs in less than 100 parallel iterations.

Of course the problem of gait learning may appear as too simple because the solution found is obvious and does not need a computer to obtain it. Nevertheless, the good and simple results obtained open the door to a most interesting study concerning working in a damaged state: if the robot can learn autonomously to walk, it could also be able to learn to walk with, for example a broken leg. This work is presented in Section 5.4.

5.3 Obstacle Avoidance Learning

The global behaviour termed as "obstacle avoidance" can be translated in measurable tasks as:

• Not to fall down.
• Not to remain immobile.
• No obstacle in a "safety" space region around the robot.
• No to go back straight away (in order to not redo infinitively the same movement: go back then advance and so on).

The critic function can then be described using these tasks or in an abbreviate formulation: when an obstacle is encountered: go back and turn. The work centred on obstacle avoidance was essentially done using the software platform. The obstacle detection is defined using a fictitious half of sphere at the head of the robot in which no obstacle can stay (as could be performed by an ultrasonic sensor).

5.3.1 Hierarchisation of the Behaviours

To avoid obstacles, a robot needs to be able to walk. Two behaviours are then to be learnt. We therefore decided to use a preprogrammed behaviour selection. At the beginning of each experiment, the robot uses the gait learning network. As soon as an obstacle is detected in front of it, the second network takes command, and remains active as long as the robot is avoiding the obstacle. When no more obstacle is present in the safety zone (ten steps without obstacle), the first network prevails again.

5.3.2 Global Network Architecture

The network devoted to control the obstacle avoidance is similar to the one used for gait learning. It is composed of six neurons, each one for one leg. The inputs are also the leg's position.

As presented previously, the solution using only one more complicated network computing the two tasks of gait learning and obstacle avoidance leads to a more difficult problem for two reasons:

• The single network becomes more complicated than the two simpler networks, so the search in the space of the coefficients would be more difficult.
• The binary information coming from the environment is different for the two tasks, so it make no sense to apply it to the same coefficients. Technically, it also make no sense to superpose on the same coefficients the behaviours learned with different parameters u and β.

This architecture is moreover warmly recommended in a related paper, for it leads to an increases in robustness, flexibility and convergence rapidity [16].

One global simulation has been done including the two networks and the two responses coming from environment (Figure 10). The switch from one network to the other is triggered by the sensing or the vanishing of an obstacle. In this case, the behaviour of the both networks are strictly identical to the behaviours obtained separately. The choice of the initialisation of the leg s position when switching from a network to another was not obvious. Drawing inspiration from animal motion, we initialise the network, for its first iteration, at a constant set of values leading to a stable position.

5.3.3 Simulations

Trajectories
Simulations have been done on this problem and the results are satisfactory. When the simulated robot meets an obstacle, it tries first to go left and to go right randomly. After this first stage the simulated robot adopts a "strategy" and chooses between left and right. This choice leads it to avoid the obstacle.

As for learning how to walk, a study was performed in order to determine the right choice of the parameters. It appears that the conclusions are qualitatively the same: $\mu+$ has to be greater than $\mu-$, and β must have a medium value. Finally, the work described below was performed taking: $\beta = 3$, $\mu+ = 0.5$ and $\mu- = 0.00001$. The time taken for evasive action is generally small: it can be 10 parallel iterations, the mean is estimated at 60 iterations including the iterations without movement (proposed actions filtered by the "muscular filters"). We estimated that the evasive process was achieved when the robot performed ten steps without encountering an obstacle; this value was in agreement with the visual estimation of the trajectories. Simulations also show that the number of falls during the avoidance is not too great: depending on the relative position of the obstacle we estimate it between 5 and 15 falls for 60 iterations before avoidance. Estimations are performed on 20 trials with identical initial conditions (only the stochastic part is changing).

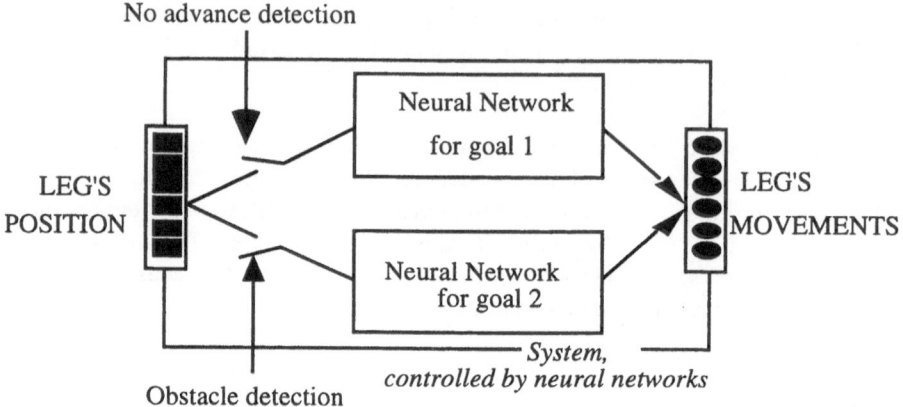

Figure 10. The global network architecture.
The global architecture includes the processes of both gait and obstacle avoidance learning. The two tasks are implemented by two networks. Only one network is active at a time, the choice of the active network is triggered by the information coming in from the environment (obstacle or not).

Figure 11 gives an example of two trajectories: when the robot meets an obstacle it solicitates the "obstacle network" which leads it to go back in turning. The obstacle having disappeared, the robot starts in the direction previously established with the "walk network". Two possibilities are then possible: if the obstacle is no longer in the sensitive area, the robot walks straight away; on the other hand, if the robot detects the obstacle once again, the "obstacle network" is then re-activated and leads the robot to turn again. Several such cycles are achieved by the robot, they appear in the figure as small straight lines taking different orientations.

Avoidance learning or continuous adaptation ?
All simulations always leads to obstacle avoidance, however, measuring if the robot effectively capitalises its experiences and learns something is not easy to do because of several factors which make the convergence of learning non repeatable:

• The stochastic part of the learning algorithm.
• The variability of the position, and of the view angle of the robot relative to the
        obstacle.
• The various dimensions and shapes of the obstacles.

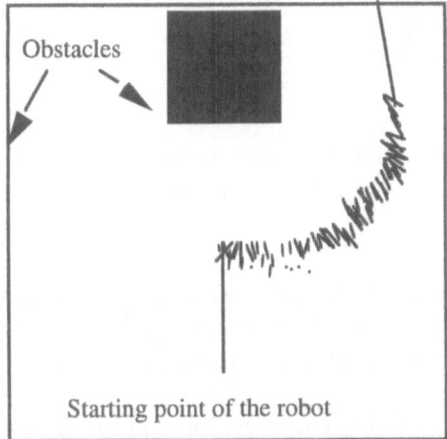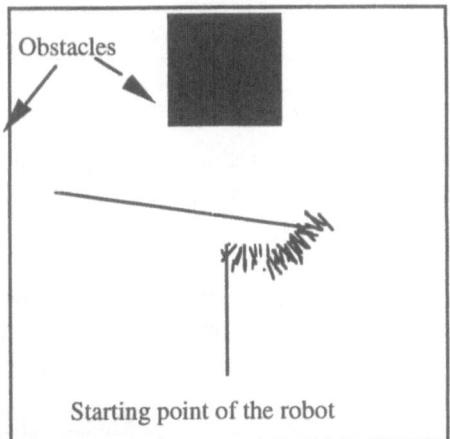

Figure 11. Obstacle avoidance: examples of trajectories.

The behaviour of obstacle avoidance is more difficult to appreciate than that related to gait learning because of the increased complexity of the robot's response to the task. We can observe that the simulated robot avoids obstacles but the way to measure its effective learning is not obvious. Moreover it would be interesting to know if the robot learns the task, or if it always adapts itself to a new configuration. The next paragraph focuses on this question.

Nevertheless, despite the lack of plausibility of the situation in an "unpredictable environment", we repeated the learning sequence on a fixed identical obstacle in order to measure the ability of the network to learn the avoidance behaviour. In this way, it is possible, for example, to estimate the experience capitalised by the robot in looking at the coefficients of the neural network.

The observations show that generally, the coefficients converge towards a limit value during an avoidance; but this does not necessarily mean that the robot adopts a particular trajectory defined by this set of limit values. This very interesting question can be tackled in observing the values of the set of synaptic coefficients, during a repetitive learning faced with the same obstacle and without reinitialisation of the coefficients. The objective is then:

• To verify if the coefficients effectively converge during this repetitive obstacle avoidance (OA).
• To associate, if possible, the convergence to a particular trajectory.

The criteria we used for measuring the convergence of the set of coefficients during the succession of OA is based on the quadratic difference between the coefficients for two consecutive OA. We can denote this measure of convergence as Mc. Then:

$$Mc(oa) = \sqrt{\sum_{ij} \frac{(c_{ij}(oa + 1) - c_{ij}(oa))^2}{n_{coeff}}} \, ,$$

where oa is a variable denoting the number of OA, $c_{ij}(oa)$ is the value of the coefficient c_{ij} after the avoidance of the obstacle oa, n_{coeff} is the number of synaptic coefficients.

In the particular case considered, the network has 36 coefficients, and the presentation of the obstacle is repeated 500 times. The results obtained on 30 simulations can be interpreted on two levels: the convergence, and the trajectory.

Firstly it appears that the coefficients always roughly converge. More precisely it is possible to distinguish 3 families of convergence. The three classes appear more explicitly when plotting the criteria of convergence versus the number of OA as shown in Figure 12.

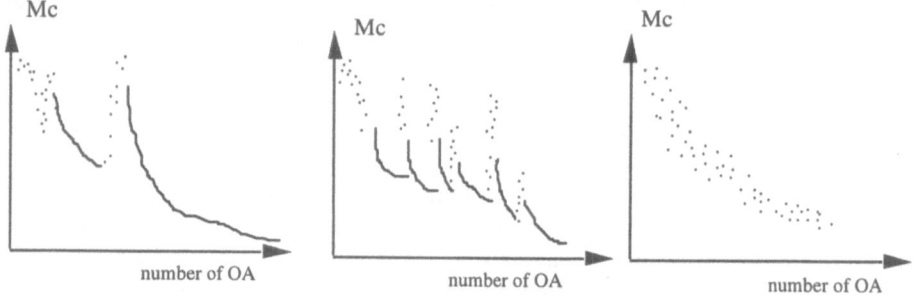

Figure 12: Three types of convergence during obstacle avoidance.
The measure of the convergence of the synaptic matrix towards a limit value is evaluated by Mc: the quadratic difference between two matrices at the end of two consecutive obstacle avoidances (OA). Mc is plotted versus the number of OA. Principally three types of convergence family can be observed: on the left, the regularity of the curves suggests that the learning is effective, in the center there is more difficulty to find a final solution, on the right, the convergence occurs, but through a cloud of values. In this last case it is possible that the device was always adapting without performing a real memorisation.

- First class: the beginning of the curve is a cloud of values in which the noise is preponderant; after 50 or 100 OA, curves appear describing a clear convergence of the set of coefficients towards a minimum , breaks can occur, but the regularity of the curves suggests that the network has effectively learned something.
- Second class: as for the first class, the beginning of the figure shows a noise preponderant evolution; afterwards some curves appear also but in greater quantity suggesting that the breaks induced by the noise add too much

confusion to the learning. Because of the presence of several curves the convergence generally appears later than the first class.
• Third class: there is roughly a convergence, but across a cloud of values globally decreasing as the curve of the first class evolution.

Plotting the number of necessary iterations to avoid the obstacle in one particular OA, versus the number of OA, as shown in Figure 13, allows some correlation with the curves previously presented to be established (same abscissa). It appears that the clear curves of the previous representation correspond to a constant number of iterations needed to avoid the obstacle. Moreover, after an examination, on the one hand the number of OA, and on the other hand the trajectory, it appears that a clear curve corresponds effectively to the same trajectory: each point of the curve is an OA, and each one has the same trajectory.

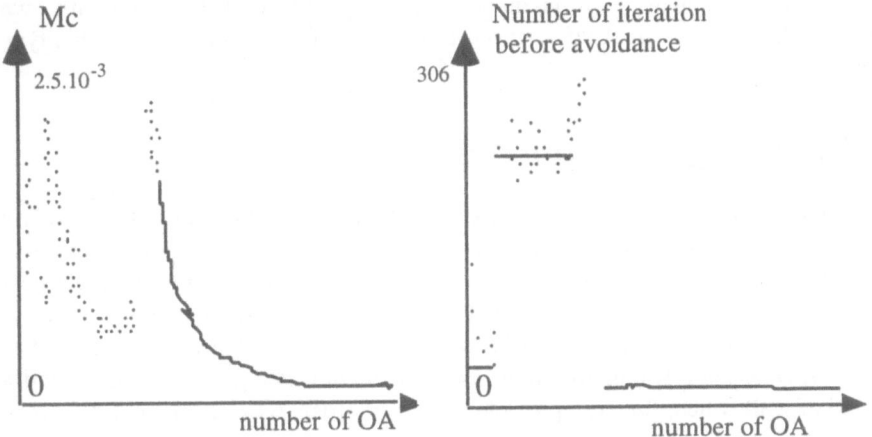

Figure 13. Correspondence between the curves of Mc and the trajectories. *One can see from this figure, on the left an example of diagram as described in Figure 12. On the right, the number of iterations needed to avoid the obstacle is plotted versus the n_j of OA, for the same experiment. An obvious correspondence between the clear curves on the left and the quasi straight lines on the right shows that a curve corresponds to a constant time to avoid the obstacle. Another analysis linking this result to trajectories shows that a curve corresponds to a particular trajectory.*

The neural network has therefore learnt to avoid the obstacle, using one particular trajectory, which is not generally the optimal one. The learning can then be done, not in the common sense in optimising the time or the movement, but in memorising through its coefficients the first favourable trajectory it performed.

5.3.4 Conclusion

Two categories of conclusions can be drawn from this study:

- First, the use of two separate networks where, one or the other is active, depending on the prevailing behaviour, is a good choice. As expected, it allows one behaviour not to be forgotten when the other is learned. It allows also quick convergence and greater reliability.
- Secondly, experiments have shown that, despite the noise inherent in the method, the avoidance networks may learn to avoid the obstacle, not randomly, but in learning a particular trajectory. We have moreover verified that this trajectory was learnt each time better during the repetitive avoidance of the same obstacle.

These results suggest an interest for these methods of behaviour learning, not for the synthesis of very well controlled efficient devices, but for embedded systems far from human control, and having an imperative necessity to find solutions for themselves when faced with an incongruous situation. Particularly without always imagining a hostile or extraplanetary environment the self adaptation of a system to its own failure is a challenge of great interest; the following paragraph investigates this type of problems.

5.4 Working in Damaged State

In several situations, a faulty device must immediately be mended in order to satisfy functional, or cost-linked objectives. Nevertheless, when the systems are far from any possible human action, or are too complicated to be mended in an acceptable cost limit, it would be interesting to investigate a reinforcement-type method in order to allow the device to reorganise itself after a non-fatal failure. That is why we tried to test the ability of the learning algorithm used for the hexapod gait learning, to find a new kind of walk when, for example, a leg is out of order. As a precaution, this work was only implemented on the software platform.

5.4.1 Failure Situation

Description of the failure
The simulated failure must not be fatal to the robot, but has to prevent it from performing the walk presented in Section 5.2.1. We therefore chose to lock the leg situated in the middle of the body, in a down position. In this configuration, the robot can use this leg as a crutch.

Application to the network
As the locked leg is always in the down position, the input applied to the network is constant and equal to +1. If the locked leg is leg 1, the input e_1 is always equal to 1.

Others inputs take values depending on their respective leg variable position. At the output of the neuron 1, delivering the movement of the leg $n_i 1$, the computation is done in the same manner as for other legs.

In other respects, the function implemented by the "muscular filters" preventing the robot doing the same movement twice consecutively or having some trailing legs, must be modified in order to allow the legs to trail on the ground.

Two types of situations can be observed: when the robot begins the learning with the failure or when the failure occurs after learning how to walk. The first observation is that, on approximately twenty trials for each configuration, a solution allowing the robot to walk was always found, the locked leg acting as a crutch. This behaviour can be understood in looking at synaptic coefficients' evolution.

5.4.2 Analysis of the Coefficients' Evolution

Two types of coefficients can be observed, those linked to healthy legs, and those linked to the locked leg. As expected, it appeared that the evolutions were quite different though they also have a common interpretation.

If we look at the coefficients which are not connected to the locked leg, their evolution were similar to those obtained with 6 healthy legs: after a search in the state space (moving legs randomly), a walk is found, the convergence occurs generally with large movements on the coefficients' evolution through time.

The robot learns how to walk with the failure from the beginning.
The evolutions of the coefficients linked to the locked leg versus the number of iterations are reported in Figure 14. One can note that, at the beginning, the coefficient values are low because of their initialisation. After a period of search in the state space, the learning finds an appropriate movement (large swings) after which the coefficients stabilised themselves to near zero. It can be noticed that all the coefficients have this type of behaviour, except the coefficient c_{11} which links the input of the locked leg to its output. This coefficient evolves toward the locked value $c_{11} = 1$.

The explanation we can make of this evolution is that the learning ratifies the fact that the output of the neuron 1 would not be dependant on the inputs (leg locked). The coefficients linking the output to other inputs are then disconnected (forced to 0), except the coefficient linked to the input 1. In the same way, the coefficient c_{11} is forced to 1 in order to ratify the fact that the position is always the same, and that the proposed network has to be consistent with the input of the locked position.

The failure occurs after the walk learning
When the failure occurs after learning how to walk, the evolutions of the coefficients are similar to those described in the previous paragraph. One can note from Figure 15 that after several large swings in the coefficients denoting that the learning has been achieved, the failure induces new swings at the end of which the coefficient c_{11} tends also to 1, whereas the other coefficients tend to a lower values. It is possible to note that each coefficient is not near zero as for the previous case, but the potential of all

these inputs is low and, in some sense, also disconnectsthe effect of the coefficients coming from the position of other legs and from the movement of the locked leg.

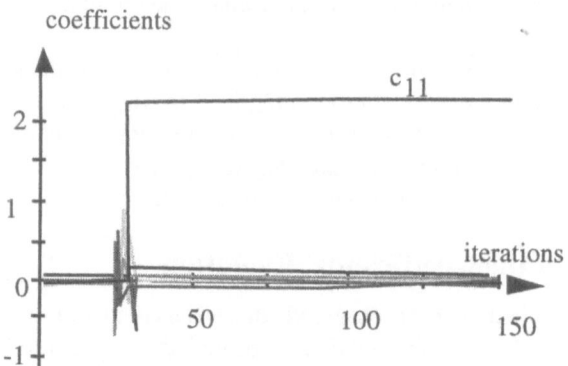

Figure 14. Evolution of the coefficients linked to the locked leg.
When the robot begins to learn with the locked leg, its behaviour is remarkable: after few iterations, all the coefficients disconnect the inputs from the output except the coefficient linking the locked position of leg to the movement of this leg. This coefficient tends to ensure that the movement corresponds to the locked position.

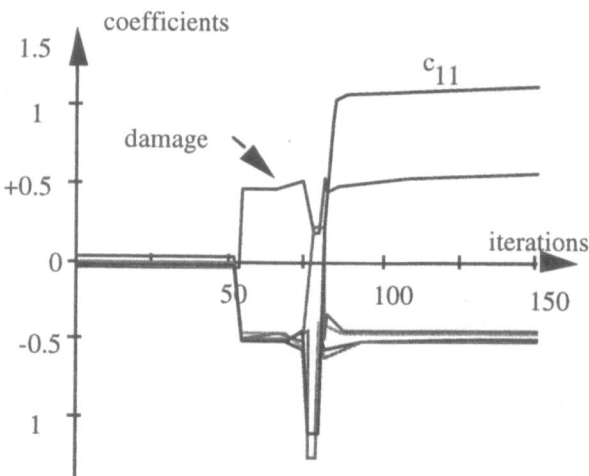

Figure 15. Evolution of the coefficients linked to the locked leg.
Walking is learned after about 50 iterations and the failure occurs approximately at 75 iterations. The locked leg is the number 1. One can then note that after large swings, all the coefficients converge. Coefficients not linked to the position of the locked leg have lower values than the coefficient expressing the link between the locked leg position and the movement which has to be deduced.

Conclusions
This new experiment shows how the property of adaptibility of the reinforcement learning can be efficient and even surprising. The capability of the method to reorganise the synchronisation of the legs in order to adapt the behaviour to a failure is really very interesting and very rich in potential applications.

6   General Conclusion

In the field of mobile robot conception,a complicated behaviour can be obtained from simple devices, and inspired by the biological example. This chapter has presented a connectionist reinforcement method in order to allow a robot to learn behaviours. Three experimental behaviours have been tested: the six legged walk, the obstacle avoidance, and the walk in damaged state. For each behaviour, the learning method converged quickly towards an interesting solution. The interpretation of the results allows us to outline two principal properties of this method. The first one is the simplicity of the concept on which it is based; the second one the efficiency of the adaptive part of the method leading to a "response time" of the order of twenty to sixty iterations. Although it is simple to make use of this method, the understanding is really more complicated due to the stochastic part of the process. For this reason the choice of simplicity of input-output coding and of network architecture was a decisive criteria of success. Moreover, the excellent results, which are only of interest if the adaptive behaviour is required *i.e.* in non stationary environment or process, can be extended to a wide range of complex or dangerous applications needing autonomous agents.

Acknowledgements

We wish to thank M. M. Artigue, M. P. Couturier, M. A. Meimouni and M. H. Silvain for their contribution to the hardware realisation of the robot.

A large team of students from the Alès School of Mines also enthusiastically took part in this work; we want here to thank M. N. Teston for learning how to walk, MM. A. Lang, E. Canal, N. Fleury and P. Ruiz for obstacle avoidance, and MM. B. Celdran and F. Favaron for learning how to walk on five legs.

Two students from the EERIE school also contributed in implementing in analog electronics the signal processing hardware. We want also to thank here MM. P. Pansier and S. Lelievre.

References

[1] Brooks, R.A. (1989), "A Robot that Walks: Emergent behaviors from carefully evolved network," *Neural Computation* 1(2), pp.253-262.

[2] Frank, A. (1988), "Walkers", *Encyclopaedia of robotics: application and automation*, DGRF R.C. ed. Wiley Interscience.

[3] Min Ming, A.C. Kak (1993), "Mobile robot navigation using neural networks and nonmetrical environment models," *IEEE Control Systems*, p.30, (October).

[4] Brooks, R.A. (1991), "Intelligence Without Representation," *Artificial Intelligence* 47, p.139.

[5] Maes, P. (1990), "A Bottom-Up Mechanism for Behavior Selection in an Artificial Creature," *Proceedings of the First International Conference on Simulation of Adaptive Behavior*, (Paris), ed. J.A. Meyer and S.W. Wilson.

[6] Beer, R.D. (1990), "Intelligence as Adaptive Behavior, an Experiment in Computational Neuroethology," *Perspective in Artificial Intelligence*, Vol. 6, Academic Press Inc.

[7] Barto, A.G., Anandan, P. (1985), "Pattern Recognition Stochastic Learning Automata," *IEEE Trans. Syst. Man Cybern.*, 15, pp.360-375.

[8] Widrow, B., Gupta, N., Maitra, S. (1973), "Punish/Reward learning with a critic in adaptive threshold systems," *IEEE Trans. Syst. Man Cybern.*, 5, pp.455-465.

[9] Bertsekas, D. (1987), *Dynamic Programming: Deterministic and Stochastic Model*, Englewood Cliffs, NJ: Prentice Hall.

[10] Cybenko, G. (1989), "Approximation by superposition of a sigmoidal function," *Math. Control Signal Systems*, Vol.2.

[11] Barto, A.G. (1996), "Reinforcement learning," *The hanbook of Brain Theory and Neural Network*, pp.804-809, Arbib edt., MIT Press.

[12] Gullapali, V. (1990), "A Stochastic Reinforcement Algorithm for Learning Real-Valued Function," *Neural Networks*, 3, pp.671-692.

[13] Sutton, R.S. (1988), "Learning to Predict By the Method of Temporal Difference," *Machine Learning*, 3, pp.9-44.

[14] Touzet, C., Sarzeaud, O. (1991), "Application d'un algorithme d'apprentissage par pénalité récompense à la génération de formes locomotrices hexapodes," *Journées de Rochebrune* AFCET (1992).

[15] Wilson, D.M. (1966), "Insect Walking," *Annual Rewiew of Entomology*, 11, pp.103-122.

[16] Long-Ji Lin (1993), *Reinforcement Learning for Robots Using Neural Networks*, PhD Thesis, Carnegie Mellon University, Pittsburgh.

Chapter 4

NEURAL NETWORKS FOR VISUAL SERVOING IN ROBOTICS

J.P. Urban, J.L. Buessler, and J. Gresser

TROP Research Group
Université de Haute-Alsace
Mulhouse
France

This chapter introduces an application of artificial neural network techniques to robotic control. Arm movements are controlled using visual features. The neurocontroller adapts on-line without any prior knowledge of the system geometry.

The neural approach is especially well suited to this kind of application. Its main limitation is the exponential increase of algorithmic cost with the number of variables. To extend the range of possible applications, a modular approach is investigated.

This modular decomposition is based on the concept of NeuroModule, and on an adaptation of the learning rules. Two combinations of NeuroModules are presented, corresponding to two variants of the proposed visual control feedback scheme.

The experiments realized both with computer simulations and on a robotic platform are very promising. They highlight both the qualities of the neural servoing scheme and the possibilities offered by the modularity of the controller.

1 Neural Networks in Robotics

1.1 Overview

Robotics opened wide-ranging research fields, with most various application perspectives. Artificial Neural Networks (ANN) have been contributing for the last 20 years to provide new robotic control approaches for robot manipulators, mobile robot navigation, as well as very specific tasks.

The synergy between robotics and ANN is especially stimulated by the need of a minimum of intelligence to control a robotic task, and by the role model offered by the animal world. The robot is a more or less autonomous system bestowed with perception and action (movement) capabilities.

Even when the robot is devised to execute repetitive tasks in an industrial environment, a minimum of autonomy, versatility, robustness in face of disturbances, is hoped for. Classical control techniques cannot offer by themselves this kind of *intelligence*. Other approaches must complement or take the place of these controllers.

The investigation fields are numerous, including adaptive control, artificial intelligence, fuzzy logic. Among those, the techniques labeled ANN present the interest of introducing some fundamental properties: a huge adaptation capability, even of self-organization with very little prior knowledge. They bestow the robot with a learning and useful information memorization capability.

Before giving an illustration of the role of learning in motor-control, it must be emphasized that the control tasks of a robot can be greatly varied. Often a hierarchy of tasks must be defined: high-level tasks (decision, trajectory planning, etc.) providing control signals to execution tasks (movement control, etc.). According to the nature of the robotic application, the elements of the hierarchy can vary greatly.

ANN find applications at all levels of this hierarchy and for the most varied tasks. The scope covered by ANN is illustrated hereafter with the topics of a few recent works: early visual processing with smart retinas [1]; ocular saccades for active vision [2]; temporal sequence elaboration [3]; self-learning truck backer-upper [4]; navigation methods of autonomous vehicles [5]; control of a walking biped robot [6]; Kendama game playing robot with human teacher [7].

1.2   Motor Control

To have a minimum of autonomy, the robot is equipped with sensors informing it continuously on its current state (proprioceptive sensors) and on its environment (exteroceptive sensors). In general, one can consider that the robot has sufficient information to perform its task effectively and to acquire some skills. The neural approach investigates ways to take advantage of this information, to determine and adapt the motor response, and to memorize the expertise acquired.

When a robot movement produces a response perceived by way of sensorial feedback, it must be emphasized that this feedback information will play a double role. First, it contributes to guide the movement, in the classical sense of control theory. But it also provides useful data to improve the control of the movement, and will be termed *learning feedback* [8].

The data used in learning feedback can be very different and lead to various learning schemes. As an example, to illustrate some of the frequently encountered difficulties, one can mention the case of a delay between action and feedback, or else the feedback can signal a failure of an action without giving any information about the correction to use (in this case *reinforcement learning* techniques can be used [8]).

Whatever the control and learning scheme may be, the key part played by the controller is to determine the mapping between an action specified in world coordinates or sometimes in sensor space, and the motor controls in actuator space. These mappings are highly nonlinear, time varying, and difficult to derive analytically.

A direct approach to neural control is to use an ANN to learn this mapping based on the motor inputs and the sensorial outputs. This supervised learning scheme, often termed *direct inverse modeling*, is simple but hardly ever sufficient (e.g. see Kawato *et al.* [9] for a discussion). It assumes in particular a one-to-one mapping between inputs and outputs and specific conditions for the learning phase.

Without specifically addressing robotic applications, Agarwal's [10] general classification of neural-network-based control approaches illustrates numerous learning combinations and applications of ANN. An excellent overview of the various neural algorithms can be found in [11].

The scheme discussed in this chapter keeps the simplicity of the direct inverse modeling approach. The velocity information in both sensor and actuator space being readily available, the Jacobian correlation can be directly learned using a supervised learning scheme.

2   Neural Visual Servoing

2.1   Introduction

Vision is certainly the royal way to endow robots with some kind of intelligence by enlarging considerably its perception capabilities. Understanding vision is a field of investigation in itself, and on the engineering side image processing techniques are the subject of numerous works and make good progress. Low-level image processing systems are presently available and represent a key component in the implementation of the approach presented in this chapter.

When machine vision provides dynamical closed-loop position control for a robot end-effector, this is referred to as *visual servoing* [12]. The approach is based on an information feedback loop which determines an error vector for every image (e.g. at video rate: 30 images/s). The error vector is defined as a measure at a given instant of the distance in image coordinates between the robot end-effector and the target position. This vector is then used to determine a new command such as to iteratively servo the error to zero.

The proposed approach (see Figure 1) shows a simple neural structure able to servo visually a robot manipulator in an efficient and robust manner. A visual servoing module, defined as a low-level task, takes care of the commands to issue to guide the robot arm to a position defined in image coordinates. This will be a vector defined in the image feature space, containing position and/or end-effector orientation coordinates, or simply image coordinates of the point to be reached.

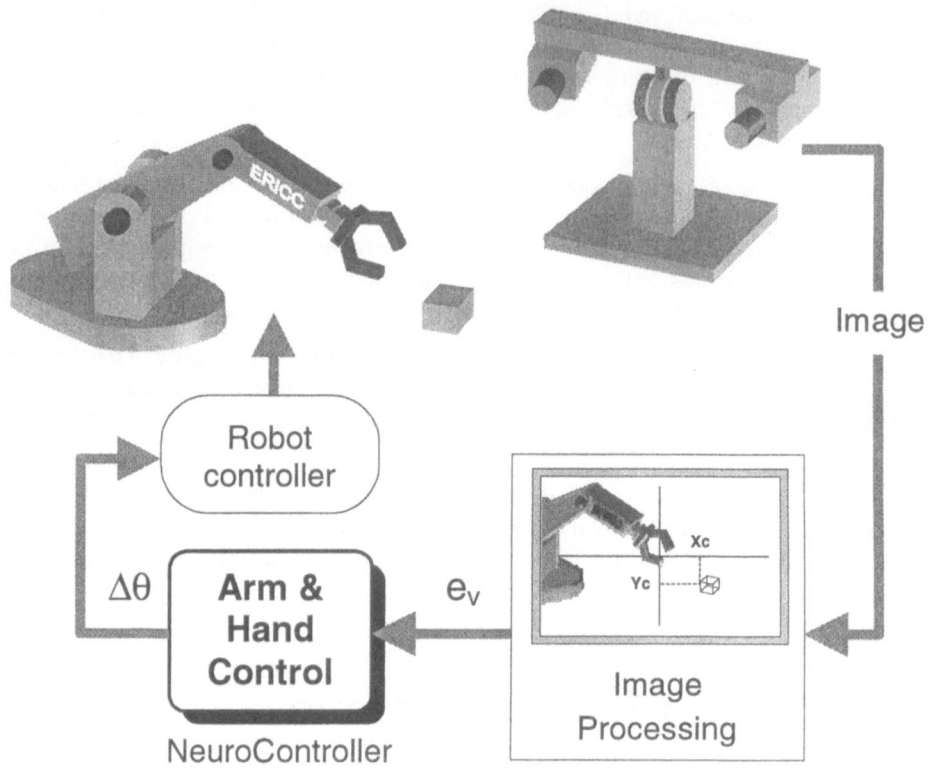

Figure 1. Overview of the visual servoing feedback loop

The robot can therefore be controlled using position commands, while other modules will determine the successive positions in order to follow a trajectory, or to reach and grasp an object, etc.

Several authors adapt the controller on-line. The Normalized Least Mean Square (NLMS) algorithm [13], known under different appellations, is very well suited to this control scheme. Jägersand [14] and Hosoda [15] showed the efficiency of this technique. The approach proposed in these pages completes this instantaneous learning with the memorization of the Jacobian relationships used in the control scheme.

This memory induces a better robustness of the algorithm and a more precise feedback control. Although the algorithmic cost is increased, the instantaneous adaptation characteristics remain: the robot-vision system is immediately usable.

The relationships on which the feedback control and the controller adaptation are based are described in the remainder of this section. The connectionist aspects will be developed in the following section.

2.2 Principle of Feedback Control

Let $\mathcal{V}$ be the space of visual features, $\mathcal{V} \subset \mathfrak{R}^m$ where m is the number of parameters describing the object. Let's consider the case of image point coordinates, expressed by vector **v**.

A continuous evaluation of the end-effector's position $\mathbf{v}_k \in \mathcal{V}$ against the target $\mathbf{c}_k \in \mathcal{V}$ allows determination of the error vector:

$$\mathbf{e}_{v,k} = \mathbf{c}_k - \mathbf{v}_k \tag{1}$$

Let **h** be the transformation function between a joint position and an image position:

$$\mathbf{h} : \Theta \rightarrow \mathcal{V} \quad , \quad \mathbf{v}_k = \mathbf{h}(\theta_k) \tag{2}$$

where Θ is the robot joint space of dimension n. The vector $\theta = \theta_k \in \Theta$ defines the robot joint position at instant k.

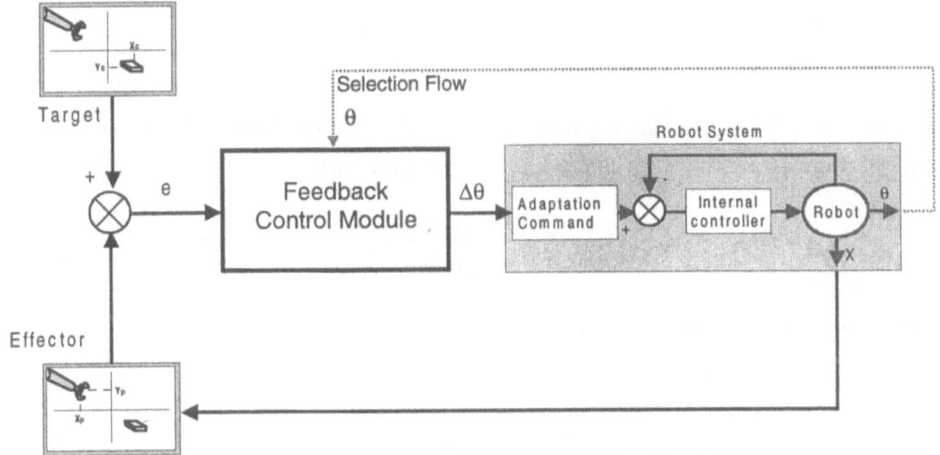

*Figure 2. Visual servoing diagram: the external visual feedback loop
is combined to a proprioceptive joint-level feedback loop*

Rather than estimating the desired joint position $\theta_c = \mathbf{h}^{-1}(\mathbf{c})$, the feedback control methods are generally based on a linear approximation in the neighborhood of the working point. The differential vector $\mathbf{d\theta}$, representing a small angular variation in joint space, is associated to differential vector $\mathbf{dv}$, representing the corresponding variation in feature space, by equation:

$$\mathbf{dv} = \mathbf{J}(\theta) \, \mathbf{d\theta} \tag{3}$$

Depending on the authors, this differential Jacobian relationship is denoted *image feature sensitivity matrix* [16] or *interaction matrix* [17].

2.2.1   Learning Feedback

The visual servoing approach requires the computation of the inverse Jacobian matrix:

$$\mathbf{J}^{-1}(\theta) = \left[\frac{\partial \theta}{\partial v} \right] \tag{4}$$

Both the robot and camera parameters, and the relative camera-robot positioning must be taken accurately into account for the exact calculation of this Jacobian. This turns out to be a difficult task, especially to ensure the validity of the data obtained from the preliminary calibration phase during the on-line computation.

The adaptive approach uses the observed robot movements. They comply with the Jacobian relationship and can be used to adapt iteratively an estimation **At** of the inverse Jacobian.

Let **dθ** and **dv** be the small displacements measured respectively in joint and image feature space at instant k. They are measured in the time interval Δt separating the sampling instants k and $(k-1)$:

$$\begin{aligned} \mathbf{dv} &= \mathbf{v}_k - \mathbf{v}_{k-1} \\ \mathbf{d\theta} &= \mathbf{\theta}_k - \mathbf{\theta}_{k-1} \end{aligned} \tag{5}$$

Let's note $\mathbf{At} \cong \mathbf{J}^{-1}(\theta_k)$, the approximation of the inverse image Jacobian. The NLMS algorithm (or delta rule) can be used to adapt **At** with:

$$\mathbf{At}_{k+1} = \mathbf{At}_k + \Delta\mathbf{At} = \mathbf{At}_k + \alpha \frac{(\mathbf{d\theta} - \mathbf{At}_k \mathbf{dv})\mathbf{dv}^{\mathrm{T}}}{\|\mathbf{dv}\|^2} \tag{6}$$

where $0 < \alpha \leq 1$ is a learning rate.

2.2.2   Feedback Control

The purpose of the control algorithm is to determine the signal to issue to the robot in order to reduce error $\mathbf{e}_v$ at each feedback cycle.

The local linear Jacobian approximation allows determination of the motor response using equation:

$$\Delta\theta = p.\mathbf{At}.\mathbf{e}_v \tag{7}$$

where $p \leq 1$ is a proportional coefficient.

This video-rate updated response ensures the applicability of the linear approximation.

3 NeuroModule and Control

3.1 NeuroModule Concept

The notion of *NeuroModule* is introduced to formalize a neural approach, and not to propose new algorithms. This approach is in the continuity of the work initiated by other groups, especially the work of Ritter *et al.* [18], broadening the scope of applications. By grouping the set of neuronal components into a single structure, the description and comprehension of the algorithms and the analysis of the proposed solutions are notably simplified.

The NeuroModule is a heterogeneous association of two neuronal algorithms, organized in two layers. A *selection layer*, based on Kohonen's Self-Organizing Maps (SOM) [19], and an *action layer*, based on the Adaline algorithm.

Kohonen's self-organizing maps are grounded on the algorithmic simplification of two neuronal mechanisms: the interaction of the neurons in the layer when they are activated (Winner-Takes-All (WTA) function), and the interaction of the neurons during the training phase (lateral plasticity control). The NeuroModule expands these two concepts to a larger number of neurons.

Without pretending to duplicate the biological reality of cortical micro-column organization [20], the image of columns of neurons is kept, and the neurons responsible for the selection mechanism are arbitrarily located in a specific layer. A selection neuron is associated to each column; therefore a Kohonen layer implementing a WTA function is associated with the set of columns. The winning neuron will activate the whole column associated with it, while the other columns are completely inhibited.

The plasticity (learning-rate factor) properties can also be extended to the notion of column. Kohonen suggests a neurophysiological interpretation: the diffusion of some special chemical transmitters in the neighborhood of an active neuron would favor the synaptic adaptation of this neuron as well as those in the neighborhood [19]. The hypothesis added here is that the modulation of the synaptic adaptation applies not only to the connections activating the selection neurons, but also, similarly, to all of the synapses of their respective columns.

One can consider that each of the columns is enticed by the same stimulus. Each column responds to the same input signals and is connected to the same output neurons. Only a single column determines the values of the layer's output, but all synapses possess locally the information essential to their adaptation.

The proposed algorithm simplifies the resulting weight adaptation:

- The learning rate is assumed constant and known for all the synapses of given column.
- The adaptation is computed for the sole active column and presented as a model to the neighboring columns.

3.2   Description and Algorithms

3.2.1   NeuroModule Response

At a given time instant, the competition between the units within the selection layer results in a winner s, the neuron whose weights best fit the *selection vector*. This neuron activates the associated column $r = s$ in the action layer:

$$\|\boldsymbol{\theta}_k - \mathbf{w_s}\| \le \|\boldsymbol{\theta}_k - \mathbf{w_r}\|, \quad \forall \, \mathbf{r} \in G \tag{8}$$

where $\mathbf{w_r}$ are the weights of the SOM layer, $\mathbf{r}$ is the column index ($\mathbf{r} \in \mathbb{N}^n$ for a selection layer of dimension n), and $\boldsymbol{\theta}_k$ is the flow selection vector at instant k.

The nature of a column can vary depending on the NeuroModule application. At first, consider that it corresponds to a matrix, and more precisely to an approximation of the inverse Jacobian matrix $\mathbf{At} \cong \mathbf{J}^{-1}(\boldsymbol{\theta}_k)$. The neural controller can be easily built.

The network's output $\mathbf{u}$ is expressed by

$$\mathbf{u} = \mathbf{At_s}\, \mathbf{e_v} \tag{9}$$

with $\mathbf{At_s}$ the weight matrix corresponding to the active column s, and $\mathbf{e_v}$ the input vector of dimension n, $\mathbf{u}$ is an m dimensional vector, with m the number of robot joints to be controlled (see Figure 3).

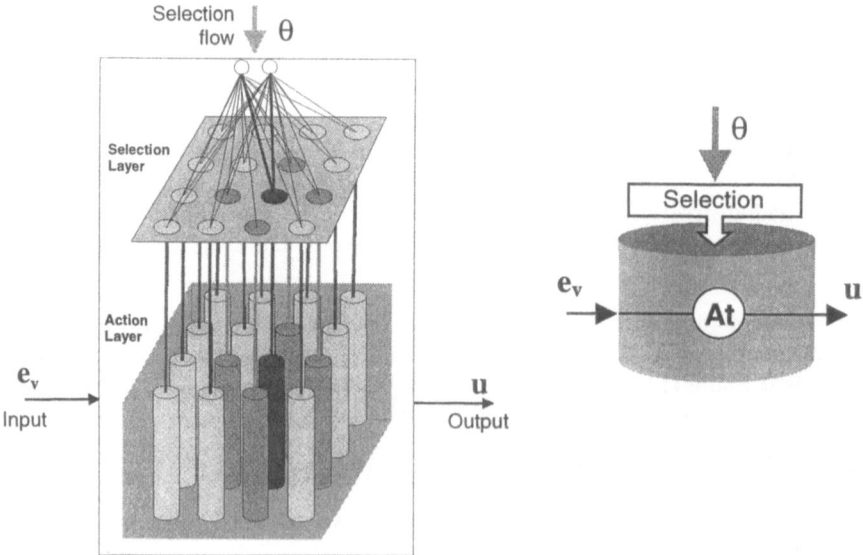

Figure 3. Functional (left) and symbolic (right) representation of the Neural Module

3.2.2 Self-Adaptation

At a given instant k, column $\mathbf{s}$ is active and the system can use the available information to adapt itself. The relationship between $d\theta = \theta_k - \theta_{k-1}$, the angular joint variation vector and $d\mathbf{v} = \mathbf{v}_k - \mathbf{v}_{k-1}$, the corresponding end-effector image position variation vector is expressed by the Jacobian relationship $d\theta = \mathbf{J}^{-1} d\mathbf{v}$.

The Adaline Network achieves a stochastic iterative estimation $\mathbf{At}$ of the Jacobian $\mathbf{J}^{-1}$ using the Normalized Least Mean Square (NLMS) rule [13]:

$$\mathbf{At} = \mathbf{At}_{\mathbf{s},k} + \alpha \frac{(d\theta - \mathbf{At}_{\mathbf{s},k} d\mathbf{v}) d\mathbf{v}^{\mathrm{T}}}{\|d\mathbf{v}\|^2} \tag{10}$$

3.2.3 Memorization

The learning activity of the columns is coordinated by the *Selection Layer*. The SOM neighborhood function is used to define a learning rate μ for each column $\mathbf{r}$ in function of the winner $\mathbf{s}$. The rate μ is the product of a term decreasing over time with a neighborhood function decreasing with the distance $\mathbf{r}$ to the winner.

For a column $\mathbf{r}$, the plasticity rate μ is a function of its distance to the active column $\mathbf{s}$:

$$\mu_{\mathbf{rs},k} = \varepsilon_k \cdot h_{\mathbf{rs},k} \quad \text{with} \quad h_{\mathbf{rs},k} = \exp\left(-\frac{\|\mathbf{r} - \mathbf{s}\|^2}{2\sigma_k^2}\right) \tag{11}$$

where the learning rate ε and the neighborhood radius σ can be constants or are decreasing over time (see e.g. [18, 19]).

The activated column $\mathbf{s}$ provides estimate $\mathbf{At}$ of the Jacobian (eqn 10) to adapt the set of columns $\mathbf{r}$ using expression:

$$\mathbf{At}_{\mathbf{r},k+1} = \mathbf{At}_{\mathbf{r},k} + \mu_{\mathbf{rs},k} (\mathbf{At} - \mathbf{At}_{\mathbf{r},k}) \tag{12}$$

3.3 Visual Servoing and the NeuroModule

The implementation of the control scheme is based on the concept of NeuroModule previously introduced. The proposed neural module will take into account the fact that the estimation of the Jacobian is local and function of the current position θ. The role of the NeuroModule is to compose, memorize, and return the coefficients necessary to the linear relationship between the error and the command. The inputs to the module are the error signal issued by the image processing system, the current joint position, and the measured image and joint velocities (Figure 4).

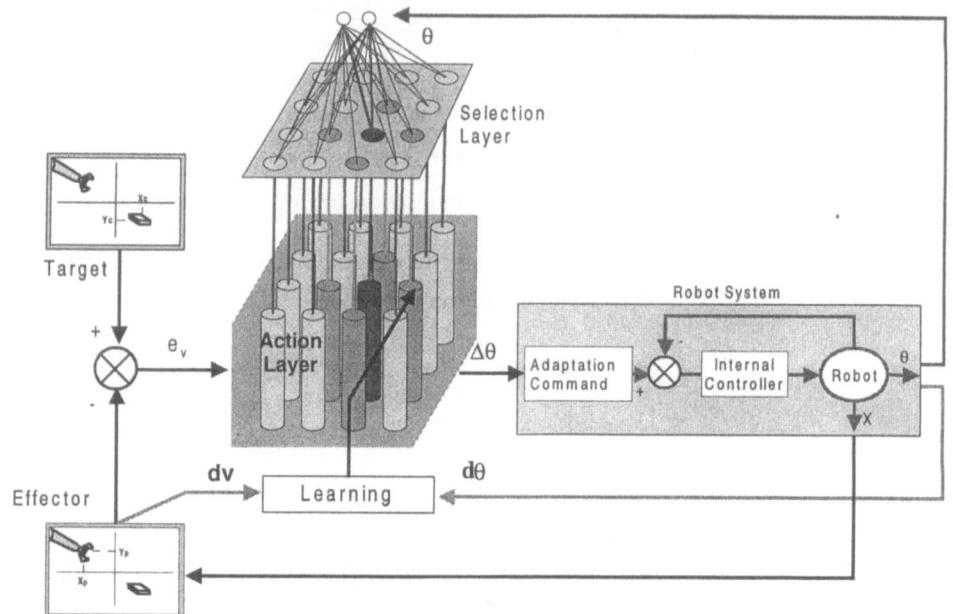

Figure 4. Visual servoing scheme with the NeuroModule

The NeuroModule has no prior knowledge about the relationships between its main input (the error signal to cancel) and the commands to issue to the robot controller. But it has been designed to acquire knowledge from its own experience: the arm movement it triggers is correlated to the corresponding image displacement. A brief preliminary probing is sufficient to gain sufficient information to move the arm on the target. The quality of the trajectories improves quickly with the ongoing movements.

The experiments on the robot platform confirm that learning is fast enough so that the module can continuously adapt to rapid changes in the geometry in the scene (e.g. a camera movement).

At each instant k, the NeuroModule goes through the following steps:

1. Activation of a column in the action layer, thus an $\mathbf{A_s}$ matrix,
2. Determination of a new estimate of the inverse Jacobian, $\mathbf{A}^* = \mathbf{A_s} + \Delta\mathbf{A}$, based on the displacement information (eqn 10),
3. Computation of response $\mathbf{u}$ (eqn 9),
4. Memorization: Adaptation of the module's neural weights (eqn 12).

3.3.1   NeuroModule Initialization

To use general experimental conditions, the columns of the action layer are randomly initialized. However, a pre-initialization could even improve the performances. The selection layer is able to adapt on-line, but the range of flow selection values is generally known. It is therefore more interesting to use this knowledge to pre-initialize the input weights to increase the neurocontroller efficiency right from the first trajectories. The results presented in this chapter are obtained using a randomly initialized action layer and a pre-organized selection layer.

3.4 Illustration: 2D Trajectory Control

A camera observes the arm displacements. The end-effector moves in a plane almost perpendicular to the optical axis of the camera. Two degrees of freedom are to be controlled.

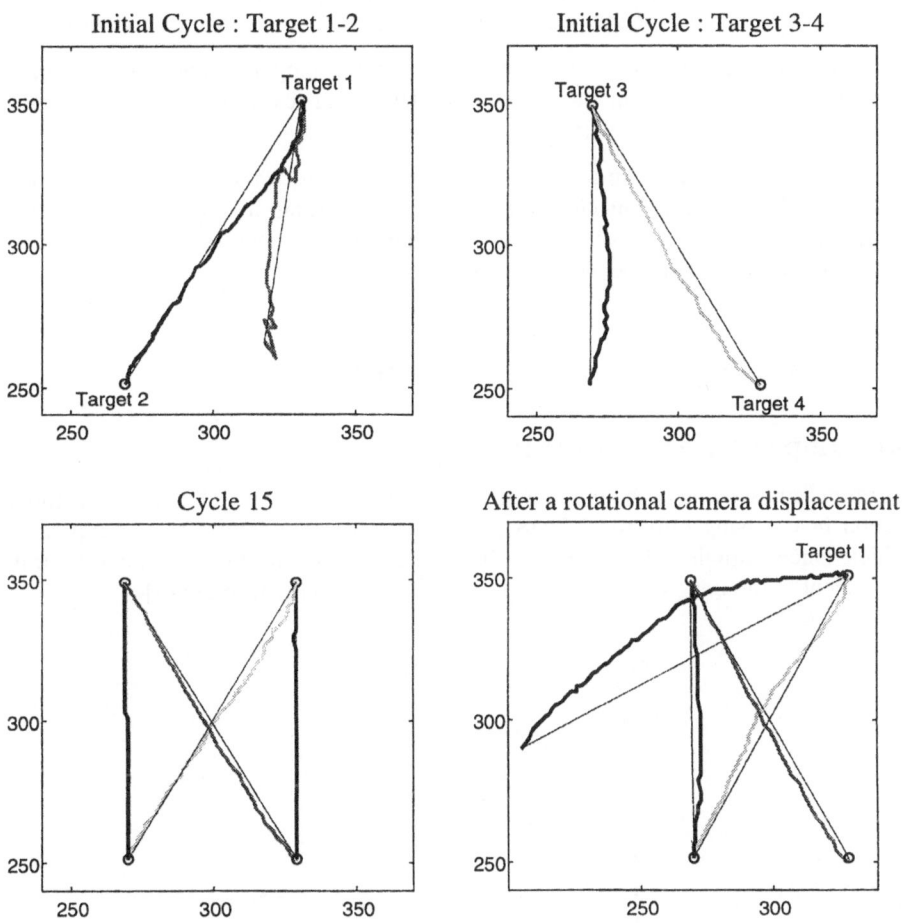

Figure 5. A 2-D robotic platform experiment:
Evolution of the trajectory in image coordinates (pixels). In the bottom right figure,
the camera is re-oriented during the movement toward target 1.

The main criteria to be met by the system are:

- The end-effector must reach the target with a defined precision (compatible with camera resolution).
- The trajectory should be as close as possible to a straight line in image coordinates.
- The duration of the movement should be as small as possible.

The neural module is comprised of 20x10 Adaline columns. Figure 5 represents the recorded end-effector movements, after a random weight initialization. A cycle is defined as the successive presentation of 4 image position targets. After some probing, the first response of the network being random, the arm reaches its first target. The following movements are smoother and more precise. Repeating the cycle of 4 targets several times, one observes a significant improvement of the trajectories. After 15 cycles, the end-effector follows almost straight lines in image coordinates.

Due to the rapid adaptation and generalization to the neighborhood, the system is operational almost right from the beginning. The initial cycle requires 3 times the time taken by the following cycles.

After 15 cycles, when the trajectories are almost linear, an abrupt change in the geometry of the system is introduced through a camera rotation. One can see that the system reacts immediately to this external change, and that the re-adaptation is fast and will ameliorate during the following cycles.

3.5   Memory and Neighborhood

3.5.1   Importance of the Memory

The proposed scheme can work without memorization, which means without selection layer, using only the NLMS algorithm. The term *MonoColumn* expresses this idea, since a single column remains in the action layer. In this case the algorithm is equivalent to the one proposed by both Jägersand [14] and Hosoda [15].

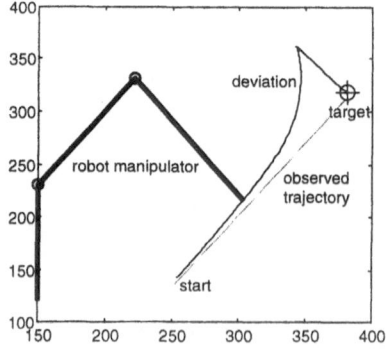

Figure 6. Case of strong deviation for a distant linear displacement command

The results obtained show that the introduction of a memory ameliorates the servoing precision. It solves an inherent problem of the algorithm. Without memory, the system adapts its unique correlation matrix **At** while the arm is moving through the workspace. It can be easily verified that this scheme fails sometimes, when the target point is at a distant location from the initial point. This problem is illustrated in Figure 6. The trajectory deviates progressively, before the system finds abruptly its ability to reach the target using a straight course. This occurs because the matrix **At** tends to become singular.

To cope with this problem, Jägersand [21] defines intermediate waypoints in the visual space. The movement toward the final target is therefore a succession of steps, each of the intermediary points being successively targets. The use of waypoints only diminishes the amplitude of the deviation. The addition of a memory avoids the formation of singularities while performing distant linear displacements.

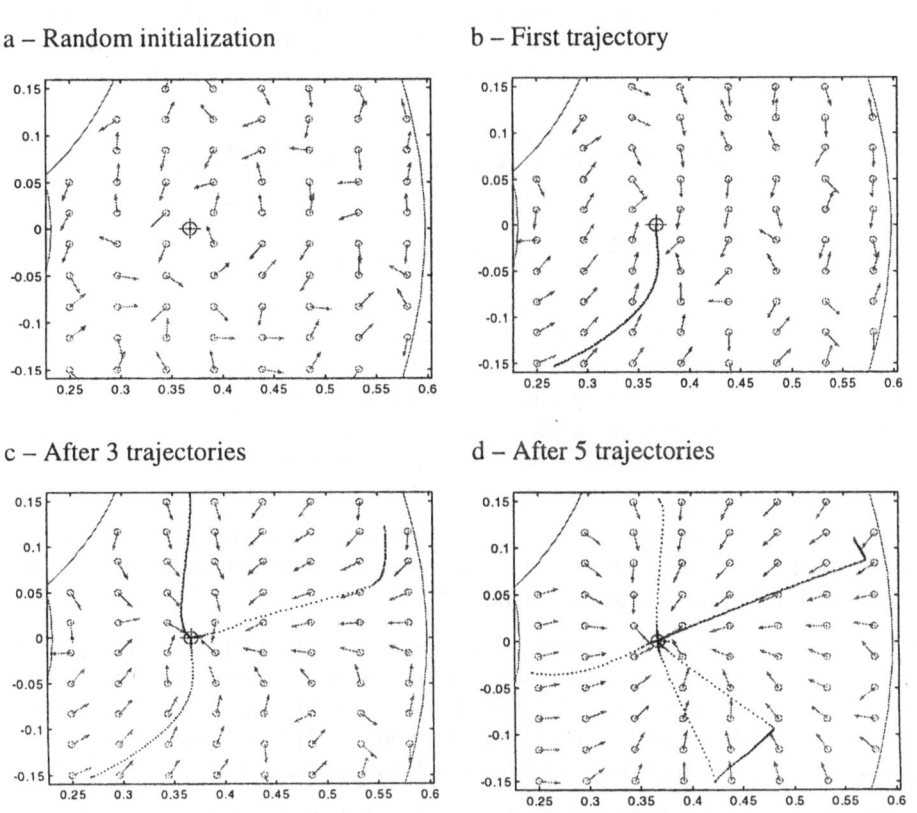

a – Random initialization

b – First trajectory

c – After 3 trajectories

d – After 5 trajectories

Figure 7. Memory with neighborhood: very rapid learning. The figures represent a set of test points (circles), a reference target (cross), and trajectories with successive positions (points). For each test point, the arrows give the direction of response to the reference target, evaluated when the trajectory is completed.

3.5.2 Neighborhood

The neighborhood function allows the rapid adaptation of a randomly initialized memory. Note that, in the case of the application considered, the memory cannot be initialized to zero values, since these initial values determine the early arm movements. In the absence of movement, learning is not possible, and the system remains in its initial state.

Figure 7 illustrates the functioning of the memory and the neighborhood. A number of test points are used, at a given instant, to evaluate the state of the memory and its ability to control its movements. The system is defined as a 2 degree of freedom planar robot arm. The response is estimated at each of the test points for the same target position. The arrow gives the direction of the first displacement toward the target. The successive points forming the trajectory are also represented.

These figures give a graphical representation of the instantaneous state of the memory and the quality of the trajectories. One can especially note the rapid amelioration of the responses, even at locations which were not directly explored by the system.

4   Cooperative Neural Modules

4.1   Introduction

When information processing is complex, it is often judicious to decompose it, for example in a succession of modules, each carrying out a partial, elementary processing. Biological brains organize themselves widely on this principle. Even for a seemingly elementary motor command like ocular saccades, Berthoz *et al.* [22] have identified several independent modules. Besides, these don't form a serial chain between perception and motor-command, but operates in multiple sensori-motor loops.

Neural techniques appropriate to modular decomposition are still barely developed (see e.g. [23] for a state of the art). Learning rules are especially difficult to derive when several interrelated modules collaborate.

Neural network applications in the field of robotics are no exception to this present limitation. Modular decompositions are nevertheless easy: numerous tasks can be managed almost independently from each other. Interesting applications where several neural modules relate mutually can be found [7, 24]. Their main limitation lies at present in the necessity to train each of the networks independently.

Modularity is nevertheless essential for more complex applications. All the more so since the works of Kawato [25] demonstrate that a neural modular approach can be much more powerful then a simple functional decomposition if it is based on a bi-directional hierarchical scheme where the information flow from one structure to the other goes systematically upward and downward. Each structure therefore provides both a direct and inverse model of the transformation achieved. Nonlinear and underdetermined problems can be solved using a rapid relaxation technique.

The scheme described hereafter is not a general solution to this line of research. However, it seems very well suited to the differential control problem. Learning of the set of modules can be realized on-line, with a simple rule and very rapidly, benefiting from bi-directional information exchanges and from the NeuroModule local linear approach.

Two examples of module combining are presented on two alternative robotic arm visual servoing tasks (see also [26, 27]).

4.2   Objective

Neurobiologist findings confirm repeatedly that the brain is not a single gigantic map, but that many modules strongly interconnected process the information in parallel and distributed manner.

Two variants of the feedback control scheme are proposed, and for each example two *a priori* plausible modules are defined. Since the memory size and the computation duration of a module depend on the number of flow selection variables, modular decomposition is interesting only if the size of the flows is reduced.

Nevertheless, these decompositions cannot lead to independent problems, and inter-module information flows are therefore necessary. Fortunately, they induce only a limited increase in memory size.

The main purpose of this work is to derive learning rules providing module adaptation. The proposed approach is efficient, quite simple, and can be easily generalized.

4.3   Parallel Module: Reaching and Grasping

This first study has been conducted on the following particular case: the end-effector is position and orientation controlled through three rotational robot axes (Figure 8). The task can be expressed as the end-effector positioning relatively to a mechanical object to be grasped. The object is placed randomly in the workspace of the robot; its elongated form assumes that the robot aligns its effector consequently to prepare for the grasping.

4.3.1   Definition of the two Modules

Works conducted in human neurobiology show that two well distinct nervous centers control in parallel the arm displacement toward an object and the pre-configuration of the hand to grasp the object. The synchronization mode between these two movements is not yet well understood [28]. We consider here these two modules in the context of a three-degree of freedom robot manipulator (see Figure 9):

- Reaching Module: controls the 2 robot axes for the space positioning of the wrist

- Pre-grasping Module: controls the configuration of the end-effector (orientation)

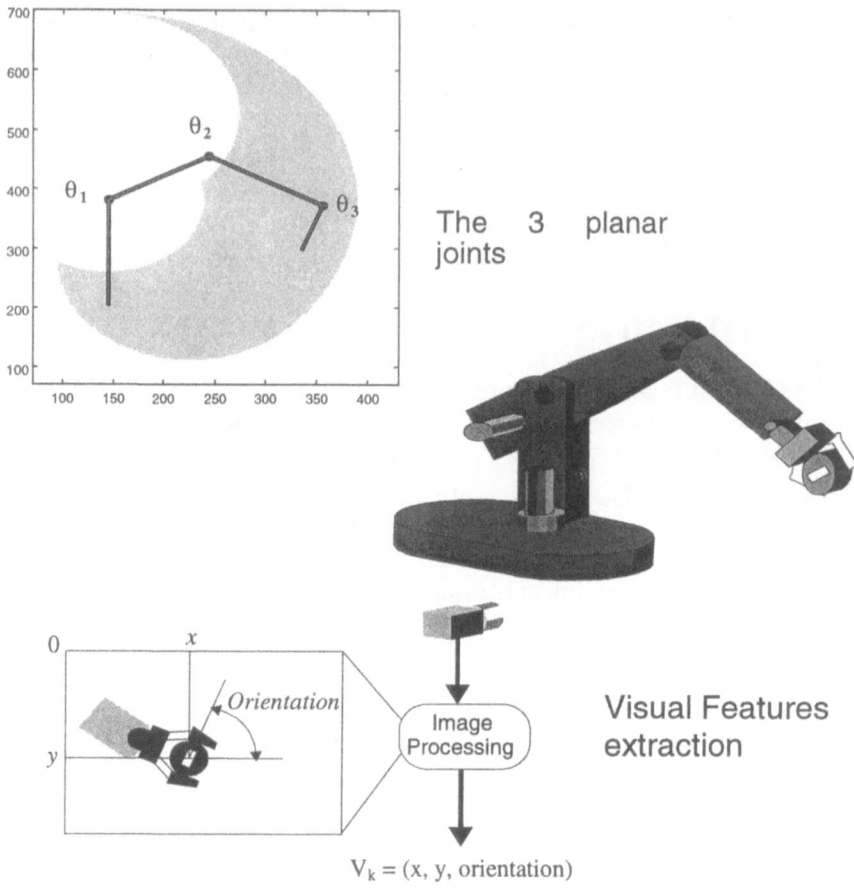

Figure 8. Reaching and Pre-Grasping task: control of 3 planar axes. The grayed area in the top left figure corresponds to the wrist's working area.

4.3.2 NeuroModule Connection

Each of the modules receives exclusively the instructions needed for its functioning and therefore realizes only a partial feedback control. However, one has to note that the two tasks are not independent. Since the image position considered is the extremity of the end-effector, to modify its orientation in keeping a constant position, all the joint positions must be modified.

In the following, we develop the approach for the control of three rotational planar axes of a robot arm, observed by a camera along a perpendicular axis.

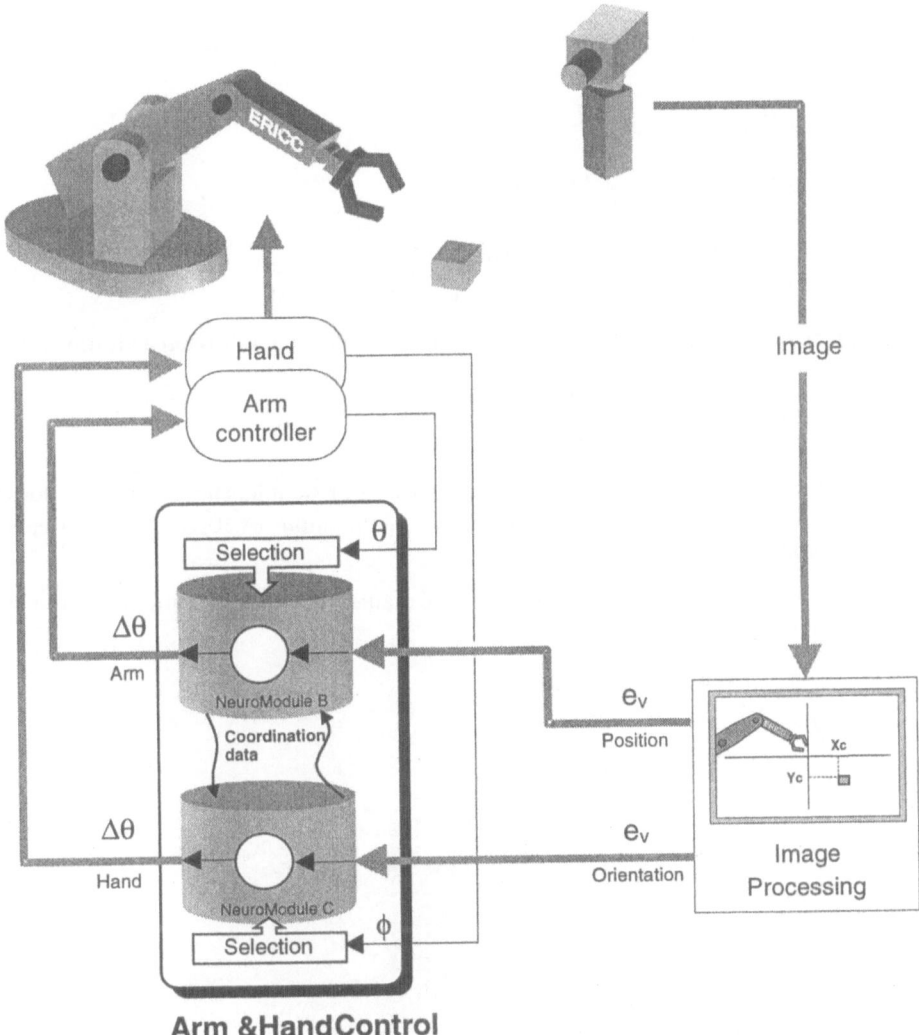

Figure 9. Visual Servoing Approach with two Parallel Cooperating Neural Modules:
organization of the data flows

The selective input vectors $\theta_{Arm} = (\theta_1,\theta_2)^T$ and $\phi_{Hand} = \theta_1+\theta_2+\theta_3$, determine the Adaline matrices of each of the modules contributing to the neural response, at a given time. These Adaline sub-networks can be described by a set of four matrices, two per module, and interconnected as shown in Figure 9.

The neural response $\Delta\theta$ to vector $\mathbf{e_v}$ can also be represented in terms of matrix products:

$$\Delta\theta_{arm} = \mathbf{Bt}(\mathbf{e}_{pos} + \mathbf{Cc}.\mathbf{e}_{orient}) \tag{13}$$

$$\Delta\theta_{hand} = \mathbf{Bc}.\mathbf{e}_{arm} + \mathbf{Ct}.\mathbf{e}_{orient} \tag{14}$$

with $\quad \mathbf{e_v} = \begin{bmatrix} \mathbf{e}_{pos} \\ \mathbf{e}_{orient} \end{bmatrix} \quad$ and $\quad \Delta\theta = \begin{bmatrix} \Delta\theta_{arm} \\ \Delta\theta_{hand} \end{bmatrix}$

The Jacobian matrix $\mathbf{At}$, here of dimension 3x3, is hence partitioned as follows:

$$\mathbf{At} = \begin{bmatrix} \mathbf{Bt} & \mathbf{BtCc} \\ \mathbf{BcBt} & \mathbf{BcBtCc} + \mathbf{Ct} \end{bmatrix} \tag{15}$$

The sub-matrices can be organized in modules. For example, since matrix $\mathbf{Bt}$ does not depend on ϕ_{hand}, it can be memorized in function of θ_{arm} in the Reaching NeuroModule.

From equation (13), one can see that the neural network is estimating the wrist position variation:

$$\mathbf{e}_{wrist} = \mathbf{e}_{pos} + \mathbf{Cc}.\mathbf{e}_{orient} \tag{16}$$

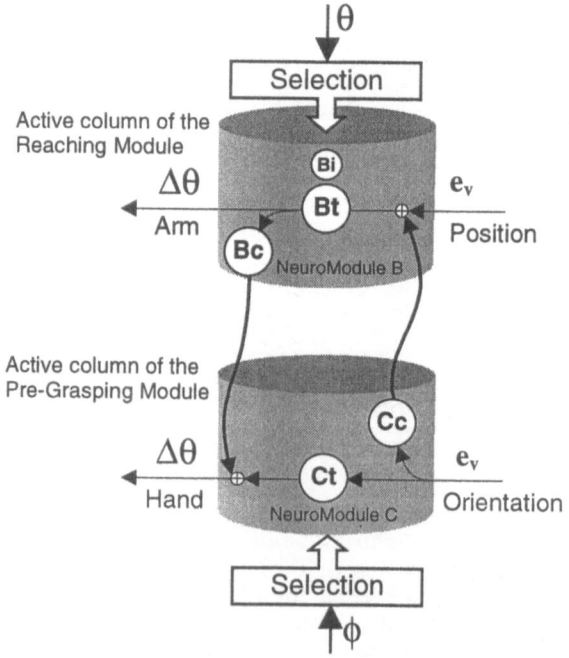

Figure 10. Cooperation of the Modules in the neural response:
Data processing in the active columns

4.3.3 Cooperative Learning

The use of neural modules or layers poses a general problem in neural networks, i.e., the evaluation of internal variables. Each module adapts its parameters (weight matrices) in function of an error obtained by comparison to a desired output. When several modules follow one another, no reference value is available to estimate the error on the internal data flows.

The classical approach is to backpropagate the final error, therefore to estimate the contribution of an internal variable to the global error measured at the output of the system.

The cooperative learning introduced in this chapter is solely based on the NLMS rule (eqn 6). All of the matrices playing a part in a partial response are taken into account. The module learns not restrictively an input-output relation, but also output-to-input relations which turn out to be helpful. In particular, a matrix **Bi** is associated with the reaching module (figure 10) to estimate e_{wrist} as the product $Bi.e_{arm}$. This matrix does not participate in the computation of the neural response, but it is needed during the weight adaptation epoch.

4.3.4 Matrix Adaptation

All matrices will be adapted in pairs. Consider for example the matrices **Cc** and **Bi**. The error can be defined as:

$$\varepsilon_1 = dv_{pos} - (Bi_{s,k}.d\theta_{arm} + Cc_{s,k}.dv_{orient}) \tag{17}$$

and the NLMS Rule is used to adapt both **Cc** and **Bi**:

$$Cc = Cc_{s,k} + \alpha_1 \frac{\varepsilon_1.dv_{orient}^T}{\left\|dv_{orient}\right\|^2 + \left\|d\theta_{arm}\right\|^2} \tag{18}$$

$$Bi = Bi_{s,k} + \alpha_2 \frac{\varepsilon_1.d\theta_{arm}^T}{\left\|dv_{orient}\right\|^2 + \left\|d\theta_{arm}\right\|^2} \tag{19}$$

The others matrices are adapted likewise. Then all matrices are memorized in their respective NeuroModules following equation (12).

The learning of bi-directional relations, with the introduction of matrix **Bi**, turns out to be an efficient technique. We expand its application field in the following application.

4.4   Serial Modules: Reaching with Active Vision

4.4.1   Active Vision Approach

In a visual servoing control scheme where the vision system acts as an observer of the scene, it is interesting to introduce active vision capabilities: automatic camera pan-tilt and zoom control will enhance the positioning precision of a robotic end-effector.

In this application the camera is mounted on a fixed robotic head. Camera orientation is obtained by two pan and tilt rotational axes. The camera is also equipped with a motorized zoom.

The camera being thus endowed with these active vision capabilities, the robotic head can be moved closer to the vertical work-plane since the camera can now go after the end-effector. The vision sensor measurements gain therefore in precision. Conversely, a static camera must be moved away from the workspace so that the end-effector is always kept in the vision field.

Two parallel closed-control loops can be distinguished in Figure 11, one regarding the vision system, and the other the robot arm. In this approach we are principally interested in the modifications to be brought into the latter one to take the mobility of the camera into account.

Note that camera rotations are servoed on the end-effector camera image. The control movements are implemented in terms of saccades (very rapid movements) to re-center the end-effector whenever it gets out of the central image zone (fovea).

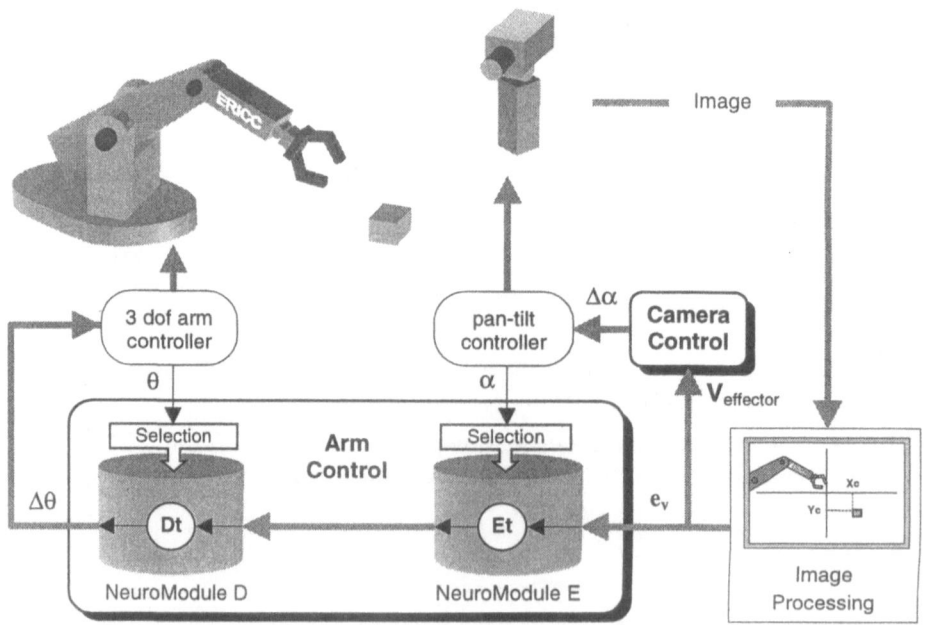

Figure 11. Synopsis of the robotic control scheme with active vision.

4.4.2   Definition of the Two Modules

The integration of active vision parameters (pan-tilt rotations) in the control scheme introduces additional control variables. Within the confined current experimentation context, these variables could be ignored. The adaptation capabilities of the neural controller are fast enough to compensate for the distortions introduced by the camera orientation changes.

However, it is more interesting to take them explicitly into account, both to enhance the performances and to prepare for a more general approach to the problem (stereoscopic vision of 3-D scenes, mobile robotic head, etc.).

This generalization will further increase the number of variables. Due to the exponential increase of memory size and processing time with the number of selection variables, the use of a single NeuroModule is unrealistic and a modular approach becomes imperative.

The Jacobian relationship can be expressed by:

$$J^{-1} = J_g^{-1} J_f^{-1} \tag{20}$$

$J_f^{-1} = J_f^{-1}(\alpha_k)$ is the inverse Jacobian of camera perspective projection function f, where α_k is the camera joint vector. $J_g^{-1} = J_g^{-1}(\theta_k)$ is the inverse Jacobian of robot direct mapping function g, where θ_k is the robot joint vector. $e_v = v_{target} - v$ is the image space error vector.

Matrix $At(\alpha_k, \theta_k)$ is an estimate of the local inverse Jacobian product. Matrix At represents the learning and memorization matrix of the neural module concept introduced hereafter.

According to equation (20) the modularization is a straightforward process. Instead of considering a single NeuroModule handling $At(\theta,\alpha)$, two modules can be chained approximating $Dt(\theta) \cong J_g^{-1}$ for the first, and $Et(\alpha) \cong J_f^{-1}$ for the second.

This decomposition is straightforward in the response mode:

$$\Delta\theta = Dt(\theta).Et(\alpha).e_v \tag{21}$$

However, the available information is not sufficient for the on-line independent learning of each of the modules: space coordinate vector $dx = J_f^{-1}.dv$ cannot be measured. The technique to solve the learning problem is briefly exposed hereafter.

4.4.3   Bi-directional Learning

The following learning scheme adapts and memorizes two matrices $Dt(\theta)$ and $Et(\alpha)$, respecting condition:

$$Dt(\theta).Et(\alpha) = J_g^{-1}(\theta).J_f^{-1}(\alpha). \tag{22}$$

Vectors $d\theta$ and dv are evaluated at each instant k and permit an iterative adaptation of the matrices based on an error between the network response and the desired value.

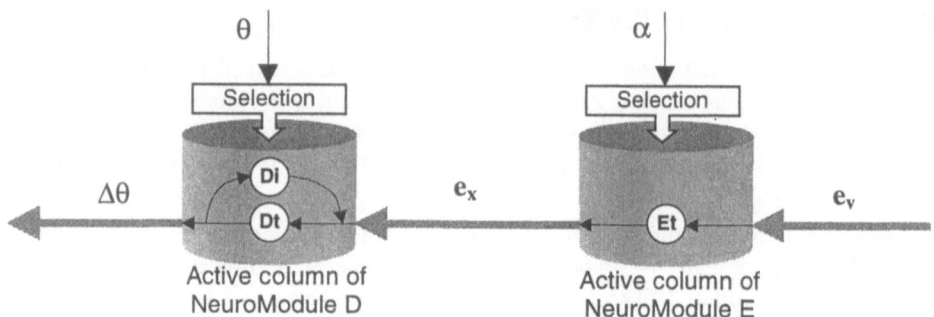

Figure 12. Module chaining: active columns and information flow

Adaptation of matrix **Dt**

Considering $d\mathbf{p} = \mathbf{Et}(\alpha)d\mathbf{v}$ an estimate of $d\mathbf{x}$, relationship (6) can be used to evaluate a new estimate of **Dt**:

$$\mathbf{Dt} = \mathbf{Dt}_{s,k} + \beta_1 \frac{(d\theta - \mathbf{Dt}_{s,k}d\mathbf{p})d\mathbf{p}^{\mathrm{T}}}{\|d\mathbf{p}\|^2} \tag{23}$$

Adaptation of matrix **Et**

To define an error for the adaptation of **Et**, we use a technique introduced in the previous section. An additional matrix $\mathbf{Di}(\theta)$ is introduced (Figure 12) and the NLMS rule is extended to this case to adapt simultaneously **Et** and **Di**:

$$\mathbf{Et} = \mathbf{Et}_{s,k} + \Delta\mathbf{Et}_k = \mathbf{Et}_{s,k} + \beta_2 \frac{(\mathbf{Et}_{s,k}.d\mathbf{v} - \mathbf{Di}_{s,k}.d\theta)d\mathbf{v}^{\mathrm{T}}}{\|d\mathbf{v}\|^2} \tag{24}$$

$$\mathbf{Di} = \mathbf{Di}_{s,k} + \Delta\mathbf{Di}_k = \mathbf{Di}_{s,k} + \beta_3 \frac{(\mathbf{Et}_{s,k}.d\mathbf{v} - \mathbf{Di}_{s,k}.d\theta)d\theta^{\mathrm{T}}}{\|d\theta\|^2} \tag{25}$$

Internal Representation Constraint

A supplementary rule is required for the adaptation of **Et**. Until now the algorithm takes only error $\varepsilon = d\theta - \mathbf{Dt}(\theta).\mathbf{Et}(\alpha).d\mathbf{v}$ into account, but doesn't fix a dimension for intermediary representation **dp**. It does therefore not prevent from drifting to zero or infinite values. We have tested a rule stabilizing internal representation $\mathbf{e}_x$. The rule forces a zero mean for weight adaptation of matrix **B**.

$$\mathbf{Et} = \mathbf{Et}_{s,k} + \Delta\mathbf{Et}_k - \beta_I.\mathbf{m}_k \tag{26}$$

with $\mathbf{m}_{k+1} = \mathbf{m}_k + \beta_M(\Delta\mathbf{Et}_k - \mathbf{m}_k)$, $\mathbf{m}_0 = 0$,

and the coefficients β_I and β_M are positive inferior to 1.

Memorization

As in the previous section, the matrices are memorized in their respective modules. NeuroModule D memorizes both matrices **Dt** and **Di**.

Two more optional criteria contribute to the rapid algorithm convergence, using the *a priori* knowledge about the learning scheme:

- Coefficients μ_2 and μ_3 are proportional to the norm of error ε.

- An NLMS rule adaptation of **Ai** imposes the correlation **Ai.A**=$\mathbf{I_d}$, where $\mathbf{I_d}$ is the identity matrix.

5   Experimental Results

5.1   General Presentation

The experiments have been conducted both on computer simulations and on a robotic platform. Many visual control feedback configurations have been tested: 3 rotational axes in 3D space or 3 co-planar axes, both with different robotic head locations, and an active scheme (pan-tilt camera) for the servoing of 2 planar axes.

Two series of results are presented hereafter, illustrating the two modular approaches introduced in Section 4.

The main experimental conditions are the following: image frequency: 25Hz, total acquisition and processing duration: 120ms (3 parallel piped stages), focal distance: 11mm, maximal angular robot axes speed: 20°/s, angular axes limits: 120°.

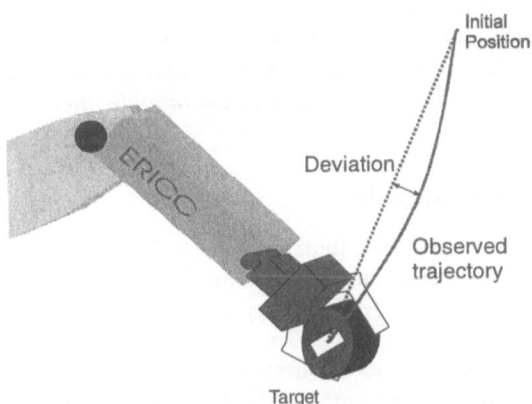

Figure 13. Trajectory evaluation using a deviation measure

5.2   Modularity Benefit - Reaching and Grasping

The neurocontroller controls the position and the orientation of the end-effector using a camera placed along a perpendicular axis (Figure 8). The experiments have been implemented at video rate (a cycle of 40 ms) with a robot-camera distance: 2 m (a pixel corresponds to approximately 2 mm at this distance). Figure 14 visualizes the observed arm trajectories.

The Selection Layer is organized during a foregoing learning phase. The results presented compare the characteristics of the visual servoing control scheme for:

- A Single Module: one module with 10x10x20 = 2000 neurons

- The cooperation of 2 Parallel Modules: 15x15 = 245 neurons for the Reaching Module and 45 neurons for the Pre-Grasping Module (a total of 290 neurons)

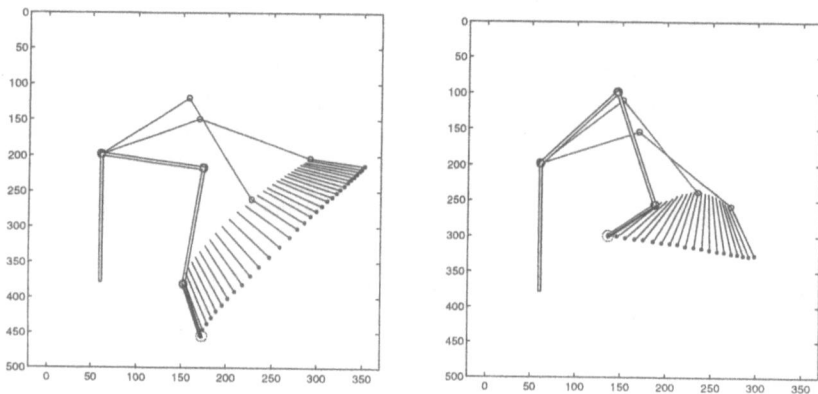

Figure 14. Illustration of the successive effector positions (camera view in pixels) while moving toward the target (only one out of two positions are drawn). The robot is represented in its initial posture, then in the middle of the trajectory (simple line), and in its final attitude (double line).

5.2.1   Simulation Results

Experiment 1: After a random initialization of the Adaline networks, a set of 100 random targets is presented to the system. This experiment has been run under different conditions (camera position, quantification of camera input, adding camera noise, image processing time, etc.). The purpose is to measure the capability of the network to adapt rapidly to its environment and to reach the first target (presented after a random initialization) within a limited time (3 times the maximum time to reach the most distant target).

The results (Table 1) show that with the exception of a few of the very early presented targets, all subsequent targets are reached.

Reached Targets	Single Module	2 Parallel Modules
First	100%	80%
2nd-10th	100%	95%
11th-100th	100%	100%

Table 1.
Capability to reach the first targets after a random initialization of the networks (number of targets reached in a limited time)

Iteration	Single Module		2 Parallel Modules	
	mean	(max)	mean	(max)
1	7.4	(53)	10.8	(24)
2	5.2	(13)	4.1	(21)
3	3.5	(10)	2.5	(7)
4	3.1	(7)	2.5	(7)
5	3.0	(7)	2.4	(7)
10-15	2.9	(7)	2.3	(7)

Table 2.
Simulation results: mean and maximum deviation (in pixels) during the repetition of a set of 20 targets. All targets have been reached in a given time

Experiment 2: In this experiment the efficiency of the learning and control realized is estimated. The ideal trajectory generated by this visual feedback control is a straight line of the end-effector position toward the target.

We measure therefore the deviation from the straight line as the farthest distance to the line (Figure 12). The experiment consists in presenting repeatedly the same set of pre-defined 20 targets.

The cooperative NeuroModules get the best performances, but they require a little longer learning time (Table 2).

5.2.2   Robotic Platform Experiment

Experiment 3: The same set of 20 targets has been used on the robotic platform (Experiment 2). Both neural approaches give similar results (Table 3). On the other hand, the computational time is considerably reduced using the coordinated modules (Table 4).

Discussion: Experiment 2 has been implemented without taking into account all the parameters of the robotic platform. The deviation measured is mainly due to the neural controller (linear approximation, learning, etc.).

In other more realistic simulations, the results get close to the ones obtained on the robot platform. The computational time and the mechanical characteristics of the robot are the main factors for the deviation. However, in all cases, the neural modules remain able to learn and to reach the targets. This is confirmed by the robotic platform results.

Iteration	Single Module		2 Parallel Modules		Table 3.
	reached targets	deviation mean (max)	reached targets	deviation mean (max)	Robotic platform: reached targets and deviation (mean and
1	85%	23 (77)	75%	23 (112)	maximum in pixels)
2	80%	13 (46)	95%	19 (76)	during the repetition of
3	100%	13 (44)	100%	19 (44)	the set of 20 targets.
4	100%	9 (40)	100%	15 (70)	
5	100%	10 (27)	100%	14 (53)	
10-15	100%	9 (30)	100%	9 (30)	

NeuroModule:	Single	2 Parallel	Table 4.
Response time	27 ms	4 ms	Computational times of the Neural Modules
Learning time (Gaussian Neighborhood)	560 ms	70 ms	on a single T805 Transputer processor.
Learning time (Bubble Neighborhood)	260 ms	38 ms	

In a similar manner, the results are affected very little by a rotation of the planar workspace in relation to the camera image plane. For example, for a 20° rotation, the convergence is still guaranteed for every target, the mean deviation increasing by half a pixel.

5.2.3   Result Summary

Both the simulation and robotic platform experiments confirm the efficiency of the cooperative approach as opposed to a single neural module:

* The resulting memory size of the cooperative NeuroModules is divided by a factor 8

* The neural response time (including adaptation and memorization) is reduced by a factor 7

* The precision obtained is equivalent; the self-adaptation and memorization properties are maintained

Each of the modules works on a subspace of the data space; it uses, provides, and memorizes coordination functions between modules.

5.3 Serial Modules - Active Vision

The experimental conditions remain unchanged, but the robotic head is now brought closer to the robotic workspace. The active pan-tilt camera is controlled with saccadic movements to maintain the robot end-effector in a 40-pixels fovea zone of the image.

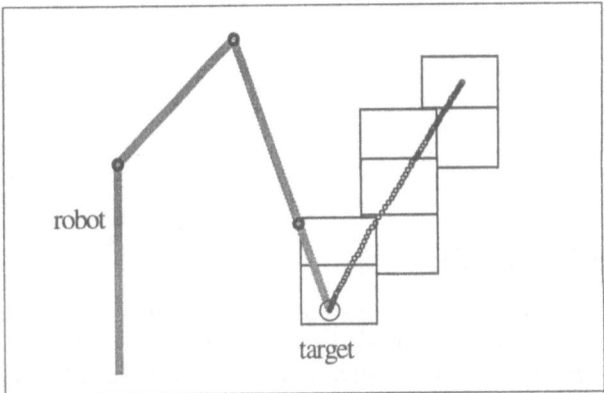

Figure 15. A trajectory with active vision. The camera follows the end-effector with rapid saccadic movements. Each rectangle represents the part of the space corresponding to the fovea. The points represent the sucessive end-effector positions.

The sizes of the modules are for NeuroModule A and D: 20x20 neurons, and for NeuroModule E: 25x15 neurons.

Figure 15 shows an end-effector trajectory toward a target position and the successive workplane areas perceived by the fovea. The controller is not disturbed by the ocular saccades.

5.3.1 Platform Experiment

Experiment 4: The robotic head is placed at a distance of 0.7 m from the robot workplane. A set of 5 targets is used which can be reached without triggering saccadic movements. By zooming in by a factor 2, active vision with saccadic control is necessary. Table 5 compares the neurocontroller's adaptation with and without saccades, where the experiments have been repeated 20 times after a random initialization.

Discussion: The feedback control precision is increased by zooming in, but it becomes necessary to foveate on the end-effector. The saccades disturbed the convergence on some targets during the first iterations. But after this, learning allows for control of the end-effector movements efficiently.

| | Without Saccades | | With Saccades | |
| | (zoom = 1) | | (zoom = 2) | |
Iteration	reached targets	deviation mean (max)	reached targets	deviation mean (max)
1	96,2%	12,1 (37,1)	86,7%	7,2 (27,5)
2	97,5%	10,8 (29,1)	93,3%	6,7 (20,1)
3	100%	8,9 (22,3)	93,3%	5,6 (16,7)
4	100%	7,5 (20,9)	93,3%	4,2 (13,9)
5	100%	7 (19,4)	100%	4 (10,2)

Table 5.
Robotic platform results: mean and maximum deviation (in mm) during the iteration (left column) of a set of 5 targets. A maximum of 12s is allotted to reach a given target.

5.3.2   Simulation Results

Experiment 5: To investigate the learning scheme more thoroughly, computer simulation experiments have been conducted. Various tests and alternatives have allowed for characterizing the functioning of the proposed algorithm. Figure 16 sums up some of these results.

Each learning series consists of 2000 targets randomly selected in the entire workspace. Like before, the experiments are repeated 20 times after initialization: the SOM networks are pre-initialized, and the Adaline columns have randomly initialized weights.

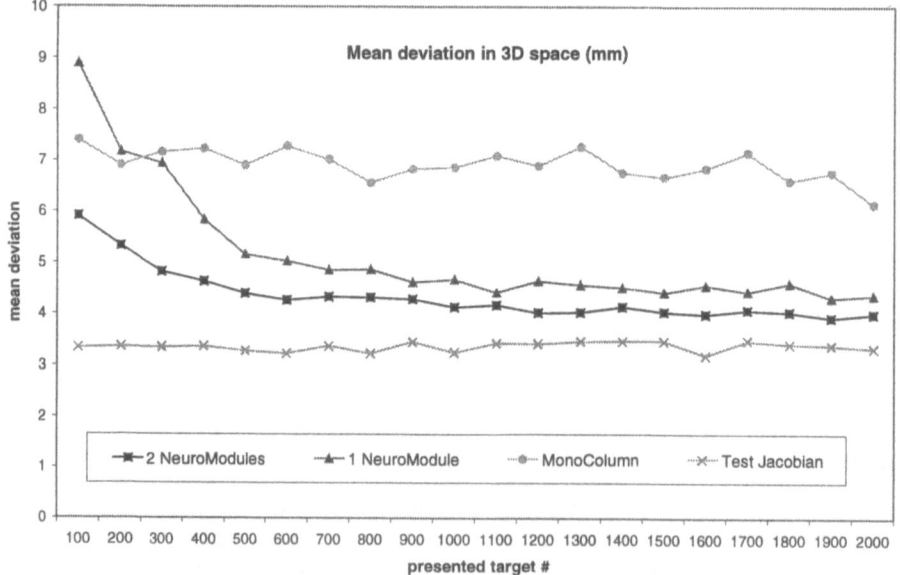

Figure 16. Trajectory precision: Comparison of 3 neurocontrollers. The means are represented for each sequence of 100 targets.

Discussion: From the first targets and onwards, the neural controller guarantees excellent performances. Several points confirm the interest of the NeuroModule's memory. The controller precision increases while it acquires experience and the results get close to the ones obtained by a precise Jacobian-based controller referred to as *test Jacobian.*

After circa 2000 targets, the neurocontroller ensures equivalent performances when learning is deactivated. However, no over-learning effects appear when learning is maintained in continuous mode, which enables on-line adaptation to any changes in the environment.

Internal Representation: The learning scheme assumes that the NeuroModules use an independent camera position representation d**p**. This representation is not supervised, but we verified that a linear relation (peculiar for each learning) allows the correlation with vector d**x** with good precision over the entire workspace (mean error of $5°$ on the orientation).

Experiment 6: The same experiments have been realized with single NeuroModule D (with only θ as information flow selection). The results are also excellent, almost equivalent to those obtained with the D-E composed NeuroModules (curve *1-Neuromodule* in Figure 15). In the former case, the on-line rapid adaptation capabilities are correcting the distortions caused by the camera rotations.

Discussion: An important point to note, since the camera is foveating on the end-effector, camera angles α are only partially independent of the robot joint positions θ. The geometry of the system and the size of the fovea will determine this plus or minus close relationship.

NeuroModule D can thus memorize most non-linearities of the Jacobian relationship in function of θ. The D-E combination of NeuroModules accommodates to this partial independence and develops even so an intermediary representation.

6   Conclusion

This chapter introduces an application of ANN techniques to robotic feedback control. Arm movements are controlled using visual features. The neurocontroller adapts on-line without prior knowledge on the geometry of the system.

The neural approach is very well suited to these kind of applications. Its main limitation being the exponential increase of the algorithmic cost with the number of variables, a modular approach has been investigated.

The modular decomposition is based on the concept of NeuroModule and on the adaptation of the learning rules. The first results, obtained both with simulation and on a robotic platform, are strongly encouraging. They have been illustrated on two kind of NeuroModule combinations, each suited to a variant of the visual feedback scheme.

Our current investigations, to be published, confirm the possibility to chain three NeuroModules with the bi-directional learning or to address more complex problems involving larger data flows. For example, the 2-D active vision experiment in Section 4.4 can be extended to the 3-D control of the arm using a binocular active-vision head.

References

[1]   Merlat, L., Silvestre, N. and Merckle, J. (1997), A Tutorial Introduction to Cellular Neural Networks, Technical Report EEA-TROP-TR-97-06, University of Mulhouse, France.

[2]   Srinivasa, N. and Sharma, R. (1997), Execution of Saccades for Active Vision Using a Neurocontroller, *IEEE Control Systems Magazine*, Vol. 17, No. 2, pp. 18-75.

[3]   Ans, B., Coiton, Y., Gilhodes, J. C. and Velay, J. L. (1994), A Neural Network Model for Temporal Sequence Learning and Motor Programming, *Neural Networks*, Vol. 7, No. 9, pp. 1461-1476.

[4]   Nguyen, D. and Widrow, B. (1991), The Truck Backer-Upper: An example of Self-Learning in Neural Networks, in: *Neural Networks for Control*, A Bradford Book, MIT Press, pp. 287-300.

[5]   Kröse, B. and van Dam, J. (1997), Neural vehicles, in: *Neural Systems for Robotics*, Academic Press, Amsterdam, pp. 271-296.

[6]   Miller, W. T. (1994), Real-Time Neural Network Control of a Biped Walking Robot, *Control Systems Magazine*, Vol. 14, No. 1, pp. 41-48.

[7]   Miyamoto, H. and et al. (1996), A Kendama Learning Robot Based on Bi-directional Theory, *Neural Networks*, Vol. 9, No. 8, pp. 1281-1302.

[8]   Barto, A. G. (1995), Reinforcement Learning in Motor Control, in: *Handbook of Brain Theory and Neural Networks*, MIT Press, pp. 809-813.

[9]   Kawato, M. (1991), Computational Schemes and Neural Network Models for Formation and Control, in: *Neural Networks for Control*, A Bradford Book, MIT Press, pp. 5-58.

[10] Agarwal, M. (1997), A Systematic Classification of Neural-Network-Based Control, *IEEE Control Systems Magazine*, Vol. 17, No. 2, pp. 75-93.

[11] Jain, A. K., Mao, J. and Mohiuddin, K. M. (1996), Artificial Neural Networks: A Tutorial, *IEEE Computer Magazine*, Vol. 29, No. 3.

[12] Hutchinson, S., Hager, G. and Corke, P. (1996), A Tutorial on Visual Servo Control, *IEEE Trans. on Robotics and Automation*, Vol. 12, No. 5, pp. 651-670.

[13] Widrow, B. and Lehr, M. A. (1990), 30 Years of Adaptive Neural Networks: Perceptron, Madaline and Backpropagation, *Proc. of the IEEE*, Vol. 78, No. 9.

[14] Jägersand, M. (1996), Visual Servoing using Trust Region Methods and Estimation of the Full Coupled Visual-Motor Jacobian, in: *IASTED Applications of Robotics and Control '96*, Applications of Robotics and Control.

[15] Hosoda, K. and Asada, M. (1994), Versatile Visual Servoing Without Knowledge of True Jacobian, in: *IROS'94*, Munich, pp. 186-193.

[16] Weiss, L. E. (1984), Dynamic Visual Servo Control of Robots : An Adaptive Image-Based Approach, Technical Report CMU-RI-TR-84-16, CMU.

[17] Espiau, B., Chaumette, F. and Rives, P. (1992), A New Approach to Visual Servoing in Robotics, *IEEE Trans. on Robotics and Automation*, Vol. 8, No. 3, pp. 313-326.

[18] Ritter, H., Martinetz, T. and Schulten, K. (1992), *Neural Computation and Self-Organizing Maps, An Introduction*, Addison-Wesley, New-York.

[19] Kohonen, T. (1995), *Self-Organizing Maps*, Springer-Verlag, Berlin.

[20] Burnod, Y. (1989), *An Adaptative Neural Network : the Cerebral Cortex*, Masson, Paris.

[21] Jägersand, M. and Nelson, R. (1994), Adaptative Differential Visual Feedback for Uncalibrated Hand-Eye Coordination and Motor Control, Technical Report TR 579, University of Rochester.

[22] Berthoz, A. and Petit, L. (1996), Looking and its Jerky Triggers of Movement, in French, *La Recherche*, No. 289, pp. 58-67.

[23] Ronco, E. and Gawthrop, P. (1995), Modular Neural Networks : A State of the Art, Technical Report CSC-95026, Center of System and Control Un. Glasgow.

[24] Littmann, E. and Ritter, H. (1993), Generalization Abilities of Cascade Network Architectures, Advances in Neural Information Processing System, Vol. 5, pp. 188-195.

[25] Miyamoto, H., Kawato, M., Setoyama, T. and Suzuki, R. (1988), Feedback Error Learning Neural Network for Trajectory Control of a Robotic Manipulator, *Neural Networks*, pp. 251-265.

[26] Urban, J. P., Buessler, J. L. and Kihl, H. (1996), Parallel Neural Processing for the Visual Servoing of A Robot Arm, in: *IEEE Int. Conf. on Systems Man and Cybernetics*, Beijing, China, Vol. 3, pp. 1806-1811.

[27] Urban, J. P., Buessler, J. L. and Wira, P. (1997), NeuroModule-Based Visual Servoing of a Robot Arm with a 2 d.o.f. Camera, to appear in: *IEEE Int. Conf. on Systems Man and Cybernetics*, Orlando.

[28] Jeannerod, M., Paulignan, Y., Mackenzie, C. and Marteniuk, R. M. (1992), Parallel Visuomotor Processing in Human Prehension Movements, in: *Control of Arm Movement in Space*, Springer-Verlag, Berlin, pp. 27-44.

Chapter 5

INTELLIGENT OPTIMAL DESIGN OF CMAC NEURAL NETWORK FOR ROBOT MANIPULATORS

Young H. Kim and **Frank L. Lewis**
Automation and Robotics Research Institute
The University of Texas at Arlington, Fort Worth, TX 76118
USA

This chapter presents the application of quadratic optimization for motion control to feedback control of robotic systems using Cerebellar Model Arithmetic Computer (CMAC) neural networks. Explicit solutions to the Hamilton-Jacobi-Bellman (H-J-B) equation for optimal control of robotic systems are found by solving an algebraic Riccati equation. It is shown how CMAC can cope with nonlinearities through optimization with no preliminary off-line learning phase required. The adaptive learning algorithm is derived from Lyapunov stability analysis, so that both system tracking stability and error convergence can be guaranteed in the closed-loop system. The filtered tracking error or critic gain and the Lyapunov function for the nonlinear analysis are derived from the user input in terms of a specified quadratic performance index. Simulation results on a two-link robot manipulator show the satisfactory performance of the proposed control schemes even in the presence of large modeling uncertainties and external disturbances.

1 Introduction

In the linear optimal control design method, a mathematical model is formulated by using knowledge of the plant dynamics. If the model equation is an accurate linear representation of the plant dynamics, one can generate an optimal control input by standard optimal control theory [10, 14]. "Feedback linearization" of nonlinear systems as a method for control design has attracted considerable interest, both in theory [29], and in such practical fields as flight control and robotics ("computed torque") [13]. The idea is to use state feedback to cancel nonlinear terms and factors followed by linear control design for the simplified system.

There has been some work related to applying optimal control techniques to the nonlinear robotic manipulator. These approaches often combine feedback linearization and optimal control techniques. Saridis and Lee [28] proposed a recursive algorithm for sequential improvement of the control law which converges to the optimal, and Luo and Saridis [17] studied linear quadratic design of PID controllers. Johansson [8] showed explicit solutions to the H-J-B equation for optimal control of robot motion and how optimal control and adaptive control may act in concert in the case of unknown or uncertain system parameters. Dawson *et al.* [6] used a general control law known as modified computed torque control (MCTC) and quadratic optimal control theory to derive a parameterized proportional-derivative (PD) form for an auxiliary input to the controller.

However, in actual situations, the robot dynamics is rarely completely known, and it is thus difficult to express real robot dynamics in exact mathematical equations or to linearize the dynamics with respect to the operating point. The real robot always includes model uncertainties such as parameter variations or nonlinearities that are not known at the time of controller design. Hence, robustness and adaptation against system uncertainties are indispensable to control system design [8].

Neural networks have been used for approximation of nonlinear systems, for classification of signals, and for associative memory. For control engineers, the approximation capability of neural networks is usually used for system identification, or identification-based control. More work is now appearing on the use of neural networks in direct closed-loop controllers that yield guaranteed performance [25, 27]. The robotic application of neural network based closed-loop control can be found [15, 18, 19]. For indirect or identification based robotic system control, several neural network and learning schemes can be found in the literature. Most of these approaches consider neural networks as very general computational models. Although a pure neural network approach without knowledge about robot dynamics may be promising, it is important to note that this approach will not be very practical due to high dimensionality of input-output space. In this way the training or off-line learning process by pure connectionist models would require a neural network of impractical size and unreasonable number of repetition cycles. The pure connectionist approach has a poor generalization properties.

The Cerebellar Model Arithmetic Computer (CMAC) [1] represents one kind of associative memory techniques, which can be used in control learning, implementing a mapping, or function approximation. It is one of the first neural networks applied to real world control problems. Its ability to locally generalize an input/output pair based on a non-linear input processing and a linear algorithm for adjusting weights guarantees fast convergence. The conventional CMAC in [1] can be viewed as a basis function network that uses local constant basis functions. For mapping and function approximation, the input space is quantized into discrete states as well as large size overlapped areas called *hypercubes*. Each hypercube covers many discrete states and is assigned a memory cell that stores information for it. The value for a quantized state is calculated as the sum of data stored in memory cells associated with the hypercubes that cover this state. While the conventional CMAC has a constant value assigned to each hypercube, the data for a quantized state are constant and the derivative information is not preserved. This problem can be solved by the modified CMAC using non-constant differentiable receptive field basis functions such as spline functions [12] and general bounded basis functions [4]. The modified CMAC is able to perform arbitrary function approximation and has differentiable receptive field basis functions. The extensive comparison with other function approximators, such as Gaussian networks and adaptive fuzzy logic system can be found in [20].

In this work, we propose a nonlinear optimal design method that integrates linear optimal control techniques and CMAC neural network learning methods. The linear optimal control has an inherent robustness against a certain range of model uncertainties [14]. However, nonlinear dynamics cannot be taken into consideration in linear optimal control design. In actual applications, it seldom happens that the plant dynamics are completely known. We thus use the CMAC neural networks to adaptively estimate nonlinear uncertainties, yielding a controller that can tolerate a wider range of uncertainties. The salient feature of this H-J-B control design is that we can use *a priori* knowledge of the plant dynamics as the system equation in the corresponding linear optimal control design. The neural network is used to improve performance in the face of unknown nonlinearities by adding nonlinear effects to the linear optimal controller.

In the neural network design, one requires both a filtered tracking error standard in robotics and a Lyapunov function [15, 26]. It is usual to select arbitrary filtered tracking errors as long as the error system is stable, and the selection of a Lyapunov function is similarly not rigorous. The importance of the approach in this work is that both the filtered tracking error or critic gain and the Lyapunov function for the nonlinear analysis are derived from the user input in terms of a specified quadratic performance index. The Lyapunov function turns out to be the value function that satisfies the H-J-B equation. The approach here should be of interest in robot manipulator control, biomechanics, flight control, and other branches of applied mechanics.

The chapter is organized as follows. In section 2 we will review some fundamentals of mathematical notations and the modified CMAC neural networks. In section 3 we give a new control design for rigid robot systems using the H-J-B equation and the relation between optimal control and stability is investigated. In section 4, a neural network controller combined with the optimal control signal is proposed. In section 5, in order to demonstrate the validity of the proposed control method, a 2-link robot controller is designed and simulated in the face of large uncertainties and external disturbances.

2 Background

2.1 Mathematical Notation

Let $\Re$ denote the real numbers, $\Re^n$ the real n-vectors, $\Re^{m \times n}$ the real $m \times n$ matrices. We define the norm of a vector $\mathbf{x} \in \Re^n$ as

$$\|\mathbf{x}\| = \left(\sum_{i=1}^{n} x_i^2 \right)^{1/2} \tag{2.1}$$

and the induced norm of a matrix $\mathbf{A} \in \Re^{m \times n}$ as

$$\|\mathbf{A}\| = \left(\lambda_{max} [\mathbf{A}^T \mathbf{A}] \right)^{1/2} \tag{2.2}$$

where $\lambda_{max} [\cdot]$ and $\lambda_{min} [\cdot]$ are the largest and the smallest eigenvalues of a matrix, respectively. The absolute value is denoted as $|\cdot|$. Also $(\cdot)^T$ denotes the transpose of a vector or a matrix.

The trace of $\mathbf{A}$, written $tr(\mathbf{A})$, is the sum of diagonal elements, and satisfies $tr(\mathbf{A}) = tr(\mathbf{A}^T)$ for a matrix $\mathbf{A} = [a_{ij}] \in \Re^{n \times n}$. For any $\mathbf{B} \in \Re^{m \times n}$ and $\mathbf{C} \in \Re^{n \times m}$, we have

$$tr(\mathbf{BC}) = tr(\mathbf{CB}) . \tag{2.3}$$

The trace of a matrix is the sum of its eigenvalues. Suppose that $\mathbf{A}$ is positive definite, then

$$tr(\mathbf{BAB}^T) \geq 0 \tag{2.4}$$

for any matrix $\mathbf{B} \in \Re^{m \times n}$, with equality iff $\mathbf{B}$ is the $m \times n$ zero matrix. Trace is a linear operator, so we have

$$dtr(\mathbf{A}) / dt = tr(d\mathbf{A} / dt) . \tag{2.5}$$

Given $\mathbf{A} = [a_{ij}]$ and $\mathbf{B} \in \Re^{m \times n}$, the Frobenius norm is defined by

$$\|\mathbf{A}\|_F^2 = tr(\mathbf{A}^T \mathbf{A}) = \sum_{ij} a_{ij}^2 \tag{2.6}$$

with $tr(\cdot)$ the trace. The associated inner product is $< \mathbf{A}, \mathbf{B} >_F = tr(\mathbf{A}^T \mathbf{B})$. The Frobenius norm is compatible with the 2-norm so that $\|\mathbf{A}\mathbf{x}\|_2 \leq \|\mathbf{A}\|_F \|\mathbf{x}\|_2$, with $\mathbf{A} \in \Re^{m \times n}$ and $\mathbf{x} \in \Re^n$.

A matrix $\mathbf{A} = [a_{ij}] \in \Re^{n \times n}$ is symmetric if $\mathbf{A} = \mathbf{A}^T$ (in other words, if $a_{ij} = a_{ji}$, $\forall i, j$). A matrix $\mathbf{A} = [a_{ij}] \in \Re^{n \times n}$ is anti-symmetric if $\mathbf{A} = -\mathbf{A}^T$ (in other words, if $a_{ij} = -a_{ji}$, $\forall i, j$).

2.2 Stability of Systems

Given $\mathbf{x} \in \Re^n$ and a nonlinear function $f(\mathbf{x}, t): \Re^n \times \Re \to \Re^n$, the differential equation

$$\dot{\mathbf{x}} = f(\mathbf{x}, t), \qquad t_0 \leq t, \qquad \mathbf{x}(t_0) = \mathbf{x}_0 \qquad (2.7)$$

has a differentiable solution $\mathbf{x}(t)$ if $f(\mathbf{x}, t)$ is continuous in $\mathbf{x}(t)$ and t.

The solution $\mathbf{x}(t)$ is said to be *Uniformly Ultimately Bounded* (*UUB*) if there exists a compact set $U \subset \Re^n$ such that, for all $\mathbf{x}(t_0) = \mathbf{x}_0 \in U$, there exists a $\delta > 0$ and a number $T(\delta, \mathbf{x}_0)$ such that $\|\mathbf{x}(t)\| < \delta$ for all $t \geq t_0 + T$ [13]. *UUB* is a notion of stability in the practical sense that is usually sufficient for the performance of closed-loop systems, provided that the bound δ is small enough.

2.3 CMAC Neural Networks

There are different ways of interpreting the CMAC neural networks. Some people decided to call it an associative memory due to its input processing and memory addressing action. Others called it a neural network, a variant of the perceptron or an interpolating memory. In fact it has features justifying all these names.

Firstly, it is a neural network, more exactly a model of the cerebellum (CMAC=Cerebellar Model Arithmetic Computer). In this function, it has been developed for the first time by Albus [1] in 1975 and used for control of a robot arm.

It has been shown, that CMAC is a special case of Rosenblatt's perceptron which is said to be the first neural network applied to more or less 'real' problems. Additionally, it is an interpolating memory, because each input point is internally represented as an association area, an approximately hypercubic environment around the input point. This means that each point to be trained into the CMAC affects not only this point itself in the input space, but also some area surrounding it.

In this work, we view the CMAC neural networks as a universal function approximator for the purpose of designing intelligent control system. In contrast to most other neural networks, where the inner states (weights) are processed in a nonlinear way, the calculations following the input processing are carried out linearly. This results in a simple (linear) training algorithm with provable convergence behavior [30].

However, the conventional CMAC with local constant basis function is not suitable for closed-loop control since its output is constant within each quantized state and the derivative information is not preserved [4]. This problem can be corrected by the modified CMAC neural networks with B-spline receptive basis function [12] or general receptive basis functions [4]. The ability of the modified CMAC neural networks to approximate continuous functions has been widely studied [4, 5, 12]. Commuri *et al.* [5] provided a rigorous mathematical analysis on the approximation property of CMAC neural network for nonlinear functions satisfying the Lipschitz nonlinearity.

Figure 1 shows the basic architecture and operation of the CMAC neural networks. The CMAC neural networks can be used to approximate a nonlinear mapping $y(\mathbf{x}): X^n \to Y^m$, where $X^n \subset \mathfrak{R}^n$ is the application specific *n*-dimensional input space and $Y^m \subset \mathfrak{R}^m$ in the application specific output space. The CMAC algorithm consists of two primary functions for determining the value of a complex function as shown in Figure 1

$$R : X \Rightarrow A$$
$$P : A \Rightarrow Y \tag{2.8}$$

where X is the continuous *n*-dimensional input space, A is an N_A dimensional association space, and Y is the *m*-dimensional output space. The function $\varphi = R(\mathbf{x})$ is fixed and maps each point in the input space onto the association space A. The function $P(\varphi)$ computes an output $y \in Y$ by projecting the association vector determined by $R(\mathbf{x})$ onto a vector of adjustable weights such that

$$y = P(\varphi) = W^T \varphi. \tag{2.9}$$

$R(\mathbf{x})$ in (2.8) is the multi-dimensional receptive field function which assigns activation values to each point in the input space X.

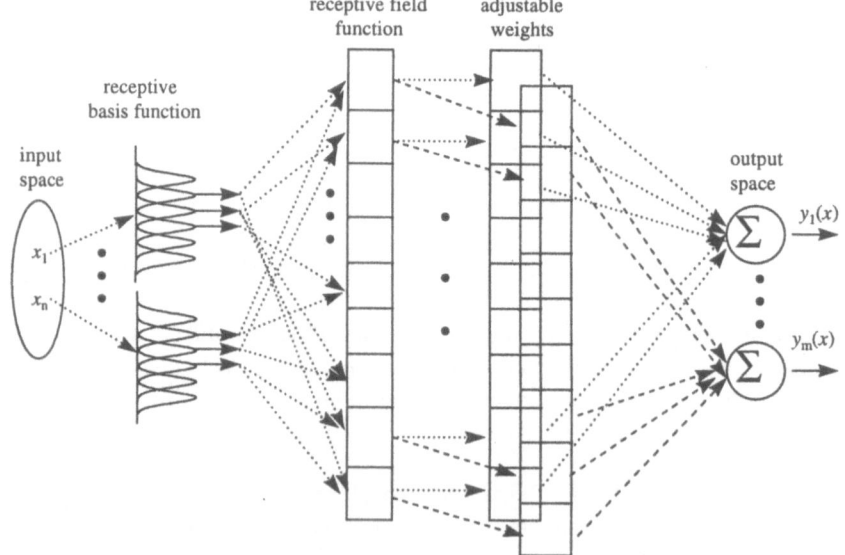

Figure 1. Architecture of a CMAC neural network.

Definition 1: Receptive field function [5, 12]

Given $\mathbf{x} = [x_1 \; x_2 \cdots x_n] \in \mathfrak{R}^n$, let $[x_{imin} \; x_{imax}] \in \mathfrak{R} \quad \forall 1 \le i \le n$ be domain of interest. For this domain select integers N_i and strictly increasing partitions

$$\pi_i = [x_{i,1} \; x_{i,2} \; \cdots \; x_{i,N_i}] \qquad\qquad \forall 1 \le i \le n.$$

For each component of the input space, the receptive field basis function can be defined as rectangular

$$\mu_{i,j}(x_i) = \begin{cases} 1 \\ 0 \end{cases} \qquad\qquad 1 \le j \le N_i \qquad (2.10)$$

or triangular

$$\mu_{i,j}(x_i) = tri(x_{i,j-1}, x_{i,j}, x_{i,j+1}) \qquad 1 \le j \le N_i \qquad (2.11)$$

or any continuous bounded function, e.g., Gaussian

$$\mu_{i,j}(x_i) = exp\left[-\frac{(x_i - m_{i,j})^2}{\sigma_{i,j}^2}\right] \qquad 1 \le j \le N_i \qquad (2.12)$$

where $m_{i,j}$ is the mean, $\sigma_{i,j}$ is the variance, and N_i is the partition number along i-th input space. The triangular function $tri(\cdot)$ is given by

$$tri(a,b,c,k) = \begin{cases} \frac{k-a}{b-a} & a \le k < b \\ \frac{c-k}{c-b} & b \le k < c \\ 0 & otherwise \end{cases} \qquad (2.13)$$

Given these one dimensional receptive field functions, multi-dimensional receptive field functions can be constructed to fully interconnect the receptive field basis functions between each input.

Definition 2: Multi-dimensional receptive field functions

Given any $\mathbf{x} = [x_1 \; x_2 \cdots x_n] \in \mathfrak{R}^n$, the multi-dimensional receptive field functions are defined as

$$\varphi_{j_1, j_2 \cdots j_n} = \frac{\mu_{1,j_1}(x_1) \cdot \mu_{2,j_2}(x_2) \cdot \cdots \cdot \mu_{n,j_n}(x_n)}{\sum_{j_n=1}^{N_n} \cdots \sum_{j_1=1}^{N_1} \prod_{i=1}^{n} \mu_{i,j_i}(x_i)} \qquad (2.14)$$

where j_i, $i = 1, \cdots, n$ represents the number of basis functions along i-th input space. It is easy to see that the multi-dimensional receptive field functions (2.14) satisfy the normal property [5, 12]. The 2-dimensional receptive field functions are shown in Figure 2 using the equally spaced Gaussian receptive basis function (2.12).

The output of the CMAC neural network is now computed by projecting this association vector onto a vector of adjustable weights w

$$y_j(x) = \sum_{i=1}^{N_A} w_{ji} \varphi_i(x) \qquad\qquad j = 1, \dots, m \qquad (2.15)$$

where $w_{ji} \in \Re$ are output layer weight values, $\varphi_i(\cdot): \Re^n \to \Re$ is the continuous multi-dimensional receptive field function and N_A is the number of association point. The effect of receptive field basis function type and partition number along each dimension on the CMAC performance has not yet been systematically studied.

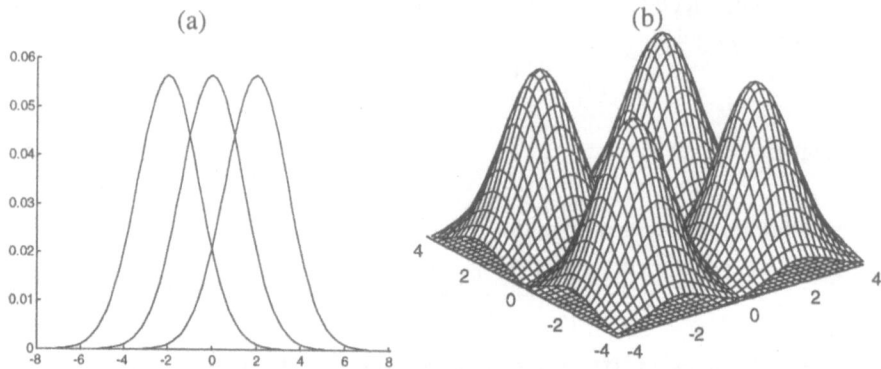

Figure 2. Receptive field shapes using Gaussian type basis function
(a) one dimension (b) two dimension.

In order to develop a stable control law we use the function approximation capability of the CMAC nonlinear network structures. To design a controller and guarantee stability, it is necessary to quantify the capability to approximate the required nonlinear functions. The CMAC offer an approach to uniformly approximate continuous functions to a specified degree of accuracy by using multi-dimensional basis functions [4, 5, 12].

The output of the CMAC can be represented as a continuously differentiable function form from $\Re^n$ to $\Re^m$ with the definition of (2.15) and expressed in a vector notation as

$$\mathbf{y}(\mathbf{x}) = \mathbf{W}^T \varphi(\mathbf{x}) \qquad (2.16)$$

where $\mathbf{W}$ is a matrix of adjustable weight values and $\varphi(\mathbf{x})$ is a vector of receptive field functions with $\mathbf{x} \in \Re^n$.

Based on the universal approximation property of CMAC, the CMAC neural networks are qualified to estimate the unknown function $f(\mathbf{x})$. In fact, there exist ideal weight values $\mathbf{W}$, so that the nonlinear function to be approximated can be represented as

$$f(\mathbf{x}) = \mathbf{W}^T \varphi(\mathbf{x}) + \varepsilon(\mathbf{x}) \qquad (2.17)$$

where $\varepsilon(\mathbf{x})$ is the *neural network functional reconstructional error*. Then for suitable neural network approximation properties, some conditions must be satisfied by $\varphi(\mathbf{x})$. It must be a *basis* [7].

In general, even given the best possible weight values, the given nonlinear function is not exactly approximated and functional reconstruction errors $\varepsilon(\mathbf{x}) \in \mathfrak{R}^m$ are remaining. We assume that an upper limit

$$\|\varepsilon(\mathbf{x})\| \le \varepsilon_{max} \qquad\qquad \forall \mathbf{x} \in \mathfrak{R}^n \qquad\qquad (2.18)$$

of the functional reconstruction error is known. For control purposes, it is assumed that the ideal approximating neural network weights exist for a specified value of ε_{max} [5, 15].

Then, and estimate of $f(\mathbf{x})$ can be given by

$$f(\mathbf{x}) = \hat{W}^T \varphi(\mathbf{x}) \qquad\qquad (2.19)$$

where $\hat{W}$ are estimates of the ideal weight values provided by adaptation learning algorithms which will be developed later. Since enabling CMAC to learn is an important issue, many researchers have focused on developing specifically designed learning rules based on the gradient descent method [7], the least square method [1], or reinforcement learning schemes [2]. In this work, the Lyapunov method is applied to derive reinforcement adaptive-learning rules for the weight values. Since these adaptive learning rules are formulated from the stability analysis of the controlled system, the system performance can be guaranteed for closed-loop control.

In particular in Barron's paper [3] it was shown that neural networks can serve as universal approximators for continuous functions more efficiently than traditional functional approximators, such as polynomials, trigonometric expansions, or splines, even though there exists a fundamental lower bound on the functional reconstruction error of order $(1/N_A)^{2/n}$. It will be shown that the effect of functional reconstructional error can be alleviated by choosing the judicious choice of fixed control gain.

2.4 Robot Arm Dynamics and Its Properties

The dynamics of an n-link robot manipulator may be expressed in the Lagrange form [13]

$$\mathbf{M}(\mathbf{q})\ddot{\mathbf{q}} + \mathbf{V}_m(\mathbf{q}, \dot{\mathbf{q}})\dot{\mathbf{q}} + \mathbf{F}_{vis}\dot{\mathbf{q}} + \mathbf{f}_{Cou}(\dot{\mathbf{q}}) + \mathbf{g}(\mathbf{q}) = \boldsymbol{\tau} \qquad (2.20)$$

with $\mathbf{q}(t) \in \mathfrak{R}^n$ joint variable, $\mathbf{M}(\mathbf{q}) \in \mathfrak{R}^{n \times n}$ inertia, $\mathbf{V}_m(\mathbf{q}, \dot{\mathbf{q}}) \in \mathfrak{R}^{n \times n}$ Coriolis/centripetal forces, $\mathbf{g}(\mathbf{q}) \in \mathfrak{R}^n$ gravitational forces, and $\mathbf{F}_{vis} \in \mathfrak{R}^{n \times n}$ diagonal matrix of viscous friction coefficients, $\mathbf{f}_{Cou}(\dot{\mathbf{q}}) \in \mathfrak{R}^n$ Coulomb friction coefficients. The external control torques to each joints are $\boldsymbol{\tau}(t) \in \mathfrak{R}^n$

Given a desired manipulator trajectory $\mathbf{q}_d(t) \in \mathfrak{R}^n$ the tracking error is

$$\mathbf{e}(t) = \mathbf{q}_d(t) - \mathbf{q}(t) \qquad\qquad (2.21)$$

and the instantaneous performance measure is defined as

$$\mathbf{r}(t) = \dot{\mathbf{e}}(t) + \Lambda \mathbf{e}(t) \tag{2.22}$$

where Λ is the constant gain matrix (not necessarily symmetric) determined later from the user specified performance index. In the derivation of CMAC adaptive learning algorithm later, the gain matrix Λ can be considered as a "reinforcement learning" [2] using nonadaptive critic, where "critic" is a measure of position displacement.

The robot dynamics (2.20) may be written in terms of the instantaneous performance measure as

$$\mathbf{M}(\mathbf{q})\dot{\mathbf{r}} = -\mathbf{V}_m(\mathbf{q},\dot{\mathbf{q}})\mathbf{r} - \tau + \mathbf{h}(\mathbf{x}) \tag{2.23}$$

where the robot nonlinear function is

$$h(\mathbf{x}) = \mathbf{M}(\mathbf{q})(\ddot{\mathbf{q}}_d + \Lambda\dot{\mathbf{e}}) + \mathbf{V}_m(\mathbf{q},\dot{\mathbf{q}})(\dot{\mathbf{q}}_d + \Lambda\mathbf{e}) + \mathbf{g}(\mathbf{q}) + \mathbf{F}_{vis}\dot{\mathbf{q}} + \mathbf{f}_{Cou}(\dot{\mathbf{q}}) \tag{2.24}$$

and, for instance,

$$\mathbf{x} = [\mathbf{e}^T, \dot{\mathbf{e}}^T, \mathbf{q}_d^T, \dot{\mathbf{q}}_d^T, \ddot{\mathbf{q}}_d^T]^T. \tag{2.25}$$

This key function $h(\mathbf{x})$ captures all the unknown dynamics of the robot arm.

Define now a control input torque as

$$\tau(t) = h(\mathbf{x}) - \mathbf{u}(t) \tag{2.26}$$

with $\mathbf{u}(t) \in \Re^n$ an auxiliary control input to be optimized later. The closed-loop system becomes

$$\mathbf{M}(\mathbf{q})\dot{\mathbf{r}}(t) = -\mathbf{V}_m(\mathbf{q},\dot{\mathbf{q}})\mathbf{r}(t) + \mathbf{u}(t). \tag{2.27}$$

This is the error system wherein the instantaneous performance measure dynamics is driven by an auxiliary control input.

The following properties of the robot dynamics are required for the subsequent development. They hold for all revolute rigid-link manipulators [13].

Property 1: *Inertia*
The inertia matrix $\mathbf{M}(\mathbf{q})$ is a positive definite symmetric matrix, and $\mathbf{M}(\mathbf{q})$ is uniformly bounded as a function of $\mathbf{q}$, i.e.

$$m_1\mathbf{I} \leq \mathbf{M}(\mathbf{q}) \leq m_2\mathbf{I} \tag{2.28}$$

where m_1 and m_2 are known positive constants that depend on the mass properties of the specific robot, and $\mathbf{I}$ is the $n \times n$ identity matrix.

Property 2: *Skew symmetry*
The matrix

$$\mathbf{N}(\mathbf{q},\dot{\mathbf{q}}) = \dot{\mathbf{M}}(\mathbf{q}) - 2\mathbf{V}_m(\mathbf{q},\dot{\mathbf{q}}) \tag{2.29}$$

is skew-symmetric. This implies $\dot{\mathbf{M}}(\mathbf{q}) = \mathbf{V}_m(\mathbf{q},\dot{\mathbf{q}}) + \mathbf{V}_m(\mathbf{q},\dot{\mathbf{q}})^T$.

3 Optimal Computed Torque Controller Design

3.1 Hamilton-Jacobi-Bellman Optimization

The motion control objective is to follow a given bounded desired trajectory $\mathbf{q}_d(t), \dot{\mathbf{q}}_d(t)$ with suitable small position errors $\mathbf{e}(t)$ and velocity errors $\dot{\mathbf{e}}(t)$. To that end, we embed the motion control problem in the following general optimization problem. Define the position error dynamics in terms of the instantaneous performance measure (2.22) as

$$\dot{\mathbf{e}}(t) = -\Lambda\mathbf{e}(t) + \mathbf{r}(t). \tag{3.1}$$

Combining (3.1) and (2.27), the following augmented system is obtained

$$\dot{\tilde{\mathbf{z}}}(t) = \begin{bmatrix} \dot{\mathbf{e}} \\ \dot{\mathbf{r}} \end{bmatrix} = \begin{bmatrix} -\Lambda & \mathbf{I} \\ \mathbf{0} & -\mathbf{M}^{-1}(\mathbf{q})\mathbf{V}_m(\mathbf{q},\dot{\mathbf{q}}) \end{bmatrix} \begin{bmatrix} \mathbf{e} \\ \mathbf{r} \end{bmatrix} + \begin{bmatrix} \mathbf{0} \\ \mathbf{M}^{-1}(\mathbf{q}) \end{bmatrix} \mathbf{u}(t) \tag{3.2}$$

or with shorter notation

$$\dot{\tilde{\mathbf{z}}}(t) = \mathbf{A}(\mathbf{q},\dot{\mathbf{q}})\tilde{\mathbf{z}}(t) + \mathbf{B}(\mathbf{q})\mathbf{u}(t) \tag{3.3}$$

with $\mathbf{A}(\mathbf{q},\dot{\mathbf{q}}) \in \Re^{2n \times 2n}$, $\mathbf{B}(\mathbf{q}) \in \Re^{2n \times n}$ and $\tilde{\mathbf{z}}(t) \in \Re^{2n \times 1}$.

A natural aim is to minimize velocity and position errors, $\dot{\mathbf{e}}(t)$ and $\mathbf{e}(t)$, with a minimum of applied torques $\tau(t)$; therefore, from the standpoint of linear quadratic control theory, it is natural to include the torques in the performance index. However, attempts to design a linear quadratic control in this manner fail due to the state-dependent nonlinear behavior of the system described in (3.3). Therefore, assuming the robot dynamics are exactly known, we formulate a quadratic performance index $J(\mathbf{u})$ as follows:

$$J(\mathbf{u}) = \int_{t_0}^{\infty} L(\tilde{\mathbf{z}}, \mathbf{u}) dt \tag{3.4}$$

with the Lagrangian

$$L(\tilde{\mathbf{z}}, \mathbf{u}) = \tfrac{1}{2}\tilde{\mathbf{z}}^T(t)\mathbf{Q}\tilde{\mathbf{z}}(t) + \tfrac{1}{2}\mathbf{u}^T(t)\mathbf{R}\mathbf{u}(t)$$

$$= \tfrac{1}{2}\begin{bmatrix} \mathbf{e}^T & \mathbf{r}^T \end{bmatrix}\begin{bmatrix} \mathbf{Q}_{11} & \mathbf{Q}_{12} \\ \mathbf{Q}_{12}^T & \mathbf{Q}_{22} \end{bmatrix}\begin{bmatrix} \mathbf{e} \\ \mathbf{r} \end{bmatrix} + \tfrac{1}{2}\mathbf{u}^T\mathbf{R}\mathbf{u}. \tag{3.5}$$

Given the performance index $J(\mathbf{u})$, the control objective is to find the auxiliary control input $\mathbf{u}(t)$ that minimizes (3.4) subject to the differential constraints imposed by (3.2). The optimal control that achieves this objective will be denoted by $\mathbf{u}*(t)$. The control variable $\mathbf{u}(t)$ is weighted by the matrix $\mathbf{R} = \mathbf{R}^T \in \Re^{n \times n} > 0$, and the vector of position and velocity errors is weighted by the matrix $Q = Q^T \in \Re^{2n \times 2n} > 0$. We will use the linear quadratic optimization process to find the optimal control $u*(t)$. It is worth noting for now that only the part of the control input to robotic system denoted by $u(t)$ in (2.26) is penalized. This is reasonable from a practical standpoint [8] since the gravity, Coriolis, and friction compensation terms in (2.24) cannot be modified by the optimal design phase.

A necessary and sufficient condition for $u*(t)$ to minimize (3.4) subject to (3.3) is that there exists a function $V = V(\tilde{z}, t)$ satisfying the H-J-B equation [10, 14]

$$\frac{\partial V(\tilde{z},t)}{\partial t} + \min_{u}[H(\tilde{z}, u, \frac{\partial V(\tilde{z},t)}{\partial \tilde{z}}, t)] = 0 \qquad (3.6)$$

where the Hamiltionian of optimization is defined as

$$H(\tilde{z}, u, \frac{\partial V(\tilde{z},t)}{\partial \tilde{z}}, t) = L(\tilde{z}, u) + \frac{\partial V(\tilde{z},t)}{\partial \tilde{z}} \dot{\tilde{z}} \qquad (3.7)$$

and $V(\tilde{z}, t)$ is referred to as the value function; it satisfies the partial differential equation

$$-\frac{\partial V(\tilde{z},t)}{\partial t} = L(\tilde{z}, u*) + \frac{\partial V(\tilde{z},t)}{\partial \tilde{z}} \dot{\tilde{z}} . \qquad (3.8)$$

The minimum is attained for the optimal control $u(t) = u*(t)$ and the Hamiltonian is then given by

$$H* = \min_{u}[L(\tilde{z}, u) + \frac{\partial V(\tilde{z},t)}{\partial \tilde{z}} \dot{\tilde{z}}] = H(\tilde{z}, u*, \frac{\partial V(\tilde{z},t)}{\partial \tilde{z}}, t) = -\frac{\partial V(\tilde{z},t)}{\partial t} . \qquad (3.9)$$

Lemma 1: The following function V composed of $\tilde{z}$, $M(q)$ and a positive symmetric matrix $K = K^T \in \Re^{n \times n}$ satisfies the H-J-B equation:

$$V = \frac{1}{2}\tilde{z}^T P(q)\tilde{z} = \frac{1}{2}\tilde{z}^T \begin{bmatrix} K & 0 \\ 0 & M(q) \end{bmatrix} \tilde{z} \qquad (3.10)$$

where K and Λ in (2.22) and (3.3) can be found from the Riccati differential equation

$$P(q)A + A^T P(q)^T - P(q)BR^{-1}B^T P(q) + \dot{P}(q) + Q = 0 . \qquad (3.11)$$

The optimal control $u*(t)$ that minimizes (3.4) subject to (3.3) is

$$u*(t) = -R^{-1}B^T P(q)\tilde{z}$$
$$= -R^{-1}r(t) \qquad (3.12)$$
$$= -R^{-1}(\dot{e}(t) + \Lambda e(t)).$$

Proof: See Appendix I.

Note that Λ is not necessarily symmetric here, in contrast to most other approaches. To proceed, the following result is required.

Lemma 2: Let the matrices $U = U^T \in \Re^{n \times n} > 0$, $C = C^T \in \Re^{n \times n} > 0$ be given. The general solution of the matrix equation

$$\Lambda^T U + U\Lambda = C \qquad (3.13)$$

with $\Lambda \in \Re^{n \times n}$ a general matrix, is given by

$$\Lambda = \Lambda_{sym} + U^{-1}Z \qquad (3.14)$$

where Z is any anti-symmetric matrix and Λ_{sym} is the unique symmetric solution to

$$\Lambda_{sym}U + U\Lambda_{sym}^T = C \qquad (3.15)$$

Proof: Let $\Lambda_1 = U^{-1}Z$ with Z anti-symmetric, then

$$U\Lambda_1 = -\Lambda_1^T U \qquad \text{or} \qquad \Lambda_1^T U + U\Lambda_1 = 0 \qquad (3.16)$$

Now add (3.15) and (3.16) to obtain

$$(\Lambda_{sym} + \Lambda_1)^T U + U(\Lambda_{sym} + \Lambda_1) = C \qquad (3.17)$$

or

$$\Lambda^T U + U\Lambda = C. \qquad (3.18)$$

Therefore Λ is a solution to (3.13). Note that Λ_{sym} has $n(n+1)/2$ elements so Λ_{sym} is a unique symmetric solution to (3.15) and Z has $n(n-1)/2$ distinct elements. So Λ has $n(n+1)/2 + n(n-1)/2 = n^2$ degrees of freedom. ∎

Notice that standard Lyapunov equation solvers (e.g. MatLab [32]) solve (3.15) for the unique Λ_{sym}.

Theorem 1: Let the symmetric weighting matrices Q, R be chosen such that

$$Q = \begin{bmatrix} Q_{11} & Q_{12} \\ Q_{12}^T & Q_{22} \end{bmatrix} > 0, \quad R^{-1} = Q_{22} \qquad (3.19)$$

with $Q_{12} + Q_{12}^T < 0$. Then the K and Λ required in Lemma 1 can be determined from the following relations:

$$K = K^T = -\tfrac{1}{2}(Q_{12} + Q_{12}^T) > 0 \qquad (3.20)$$

$$\Lambda^T K + K\Lambda = Q_{11} \qquad (3.21)$$

with (3.21) solved for Λ using Lemma 2.

Proof: See Appendix II.

Remarks:
1) Care should be taken in the selection of Q_{11}, Q_{12} and Q_{22} in order to guarantee positive definiteness of the constructed matrix Q. A way to do is to satisfy the following inequality [11]

$$\lambda_{min}(Q_{22}) > \|Q_{12}\|^2 / \lambda_{min}(Q_{11}). \qquad (3.22)$$

2) With the optimal feedback control law $u*(t)$ calculated using Theorem 1, the appropriate external torques $\tau(t)$ to apply to the robotic system (2.20) are calculated in accordance with the predefined control input (2.26), i.e.,

$$\tau*(t) = h(\mathbf{x}) - \mathbf{u}*(t) \qquad (3.23)$$

where $h(\mathbf{x})$ is given by

$$h(\mathbf{x}) = \mathbf{M}(\mathbf{q})(\ddot{\mathbf{q}}_d + \Lambda\dot{\mathbf{e}}) + \mathbf{V}_m(\mathbf{q},\dot{\mathbf{q}})(\dot{\mathbf{q}}_d + \Lambda\mathbf{e}) + \mathbf{g}(\mathbf{q}) + \mathbf{F}_{vis}\dot{\mathbf{q}} + \mathbf{f}_{Cou}(\dot{\mathbf{q}}). \qquad (3.24)$$

The external torques (3.24) will be referred to as Optimal Computed Torque Controller (OCTC) in this paper [6].

3) Compared to the control law in [8], a more general control law with less complex derivation is obtained here. There is an additional design freedom in the selection of the arbitrary anti-symmetric matrix $\mathbf{Z}$ in Lemma 2 that can be used to achieve additional performance objectives. Note that Λ is not symmetric.

3.2 Stability Analysis

It is not immediately obvious that an optimal control based on the solution (3.6)-(3.9) to the H-J-B equation does guarantee stable closed-loop behavior. Only the solutions that do guarantee a stable closed-loop behavior are of any interest for control design purposes. The next result shows that stability is guaranteed using the optimal control in Lemma 1 and Theorem 1.

Theorem 2: Suppose that matrices $\mathbf{K}$ and Λ exist which satisfies the hypotheses of *Lemma 1* and in addition there exists constants k_1 and k_2 such that $0 < k_1 < k_2 < \infty$ and the spectrum of $\mathbf{P}$ is bounded in the sense that $k_1\mathbf{I} < \mathbf{P} < k_2\mathbf{I}$ on (t_0, ∞). Then using the feedback control (3.12) in (3.3) results in the controlled non-linear system

$$\dot{\tilde{\mathbf{z}}}(t) = \{\mathbf{A}(\mathbf{q},\dot{\mathbf{q}}) - \mathbf{B}(\mathbf{q})\mathbf{R}^{-1}\mathbf{B}(\mathbf{q})^T\mathbf{P}(\mathbf{q})\}\tilde{\mathbf{z}}(t) \qquad (3.25)$$

being globally exponentially stable (GES) about the origin in $\Re^{2n}$.

Proof:
The approach taken here will be to show that if the hypotheses of *Theorem 1* are satisfied, then the value function V given in (3.10) serves as a Lyapunov function which satisfies the conditions required to guarantee global exponential stability (GES) of the system (3.25).

The quadratic function $V(\tilde{\mathbf{z}}, t)$ is a suitable Lyapunov function candidate because it is positive radially growing with $\|\tilde{\mathbf{z}}\|$. It is continuous and has a unique minimum at the origin of the error space. It remains to show that $dV/dt < 0$ for all $\|\tilde{\mathbf{z}}\| \neq 0$. From the solution of the H-J-B equation (A1.12) it follows that

$$\frac{dV(\tilde{\mathbf{z}}, t)}{dt} = -L(\tilde{\mathbf{z}}, \mathbf{u}*). \qquad (3.26)$$

Substituting (3.12) into (3.14) gives

$$\frac{dV(\tilde{\mathbf{z}}, t)}{dt} = -\tfrac{1}{2}\{\tilde{\mathbf{z}}^T\mathbf{Q}\tilde{\mathbf{z}} + (\mathbf{B}^T\mathbf{P}\tilde{\mathbf{z}})^T\mathbf{R}^{-1}(\mathbf{B}^T\mathbf{P}\tilde{\mathbf{z}})\} < 0 \qquad \forall t > 0 \quad \tilde{\mathbf{z}} \neq 0. \qquad (3.27)$$

The time derivative of the Lyapunov function is negative definite and the assertion of the theorem then follows directly from the properties of Lyapunov function [28]. ∎

4 CMAC Neural Controller Design

In actual applications, it seldom happens that the robot nonlinearities $h(\mathbf{x})$ are completely known so that computing the control torques $\tau(t)$ using (3.23) is unreasonable. Therefore, we give a neural controller design method that assumes the robot dynamics are completely unknown. We use CMAC to estimate the nonlinear uncertainties $h(\mathbf{x})$, resulting in a robust nonlinear neural controller that can tolerate a wider range of uncertainties.

The block diagram in Figure 3 shows the major components that embody the CMAC neural controller. The external control torques to the joints are composed of the optimal feedback control law given in Theorem 1 plus the CMAC neural network output components.

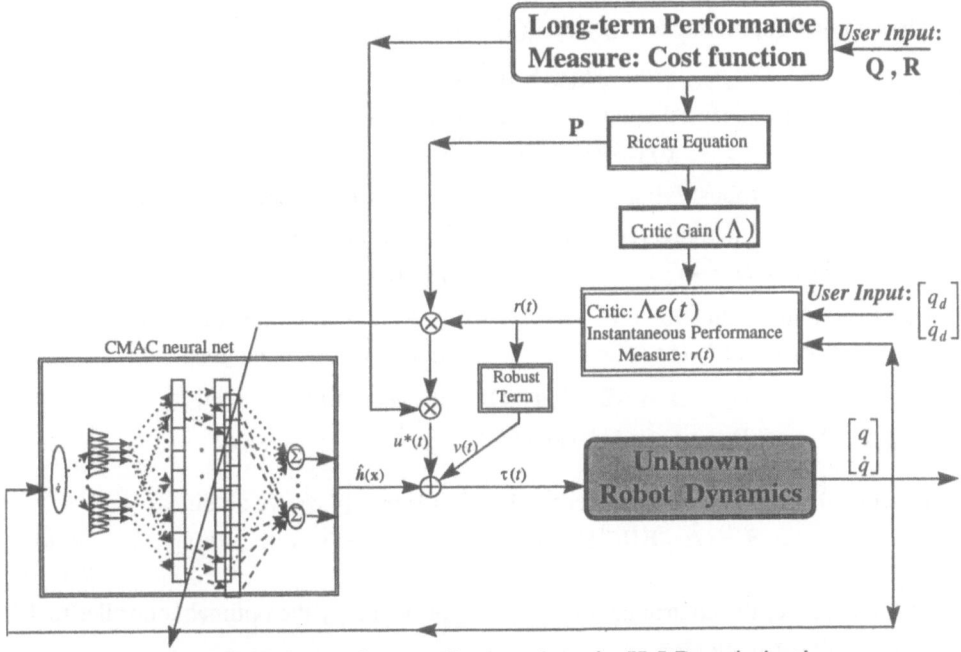

Figure 3. CMAC neural controller based on the H-J-B optimization.

The optimal feedback control law is based on errors between the desired and actual joint states. As shown later, the adaptive learning algorithm for weight values is dependent on the user specified performance index $\mathbf{Q}$, $\mathbf{R}$ through the instantaneous performance measure $\mathbf{r}(t) = \dot{\mathbf{e}} + \Lambda\mathbf{e}$. Therefore the neural network here is trained on-line to minimize the specified performance index. As the neural network learns adaptively and gradually, the feedback optimal control law contributions to the external control torques to the joints steadily diminish. Consequently the behavior of the overall closed-loop system is governed by the user specified performance index which is defined in the previous section.

Starting with the robot dynamics with the friction and external disturbance terms which are difficult to model and compensate, we have the following robot dynamics

$$\mathbf{M}(\mathbf{q})\ddot{\mathbf{q}} + \mathbf{V}_m(\mathbf{q},\dot{\mathbf{q}})\dot{\mathbf{q}} + \mathbf{F}_{vis}\dot{\mathbf{q}} + \mathbf{f}_{Cou}(\dot{\mathbf{q}}) + \mathbf{g}(\mathbf{q}) + \tau_{dis} = \tau \qquad (4.1)$$

where the unknown disturbances are denoted by $\tau_{dis} \in \Re^n$ and bounded by b_d. Note that continuous approximation to discontinuous terms in the friction force is employed. Based on the universal approximation property of the CMAC neural networks, the nonlinear robot function $h(\mathbf{x})$ in the robotic system can be represented by a CMAC neural network

$$h(\mathbf{x}) = W^T \varphi(\mathbf{x}) + \varepsilon(\mathbf{x}) \qquad \qquad \|\varepsilon(\mathbf{x})\| \le \varepsilon_{max} \qquad (4.2)$$

where $\varphi(\mathbf{x})$ is multi-dimensional receptive field function for the CMAC. Note that an continuous approximation to discontinuous terms in friction forces is employed. The *reconstruction error* $\|\varepsilon(\mathbf{x})\|$ is bounded by known constant.

Then a functional estimate $\hat{h}(\mathbf{x})$ of $h(\mathbf{x})$ can be written as

$$\hat{h}(\mathbf{x}) = \hat{W}^T \varphi(\mathbf{x}). \qquad (4.3)$$

The external torques to the joints in terms of neural network is given by

$$\tau(t) = \hat{W}^T \varphi(\mathbf{x}) - \mathbf{u}*(t) - v(t) \qquad (4.4)$$

where $v(t)$ is a robustifying vector that will be determined later in the face of external disturbances or functional reconstruction error. Then, the dynamics (2.23) becomes

$$\mathbf{M}(\mathbf{q})\dot{\mathbf{r}}(t) = -\mathbf{V}_m(\mathbf{q},\dot{\mathbf{q}})\mathbf{r}(t) + \tilde{W}^T \varphi(\mathbf{x}) + \varepsilon(\mathbf{x}) + \tau_{dis} + v(t) + \mathbf{u}*(t) \qquad (4.5)$$

with the weight estimation error $\tilde{W} = W - \hat{W}$. Note that the functional representation (4.2) of $h(\mathbf{x})$ is used. The state space description of (4.5) can be given by

$$\dot{\tilde{\mathbf{z}}} = \mathbf{A}\tilde{\mathbf{z}} + \mathbf{B}[\mathbf{u}* + \tilde{W}^T \varphi(\mathbf{x}) + \varepsilon(\mathbf{x}) + \tau_{dis} + v(t)] \qquad (4.6)$$

with $\tilde{\mathbf{z}}$, $\mathbf{A}$ and $\mathbf{B}$ given in (3.2) and (3.3).

Inserting the optimal feedback control law (3.12) into (4.6), we obtain

$$\dot{\tilde{\mathbf{z}}} = (\mathbf{A} - \mathbf{B}\mathbf{R}^{-1}\mathbf{B}^T \mathbf{P})\tilde{\mathbf{z}} + \mathbf{B}\{\tilde{W}^T \varphi(\mathbf{x}) + \varepsilon(\mathbf{x}) + \tau_{dis} + v(t)\} \qquad (4.7)$$

Theorem 3: Let the control action $\mathbf{u}*(t)$ be provided by the optimal controller (3.12) with the robustifying term given by

$$v(t) = -k_z \frac{\mathbf{r}(t)}{\|\mathbf{r}(t)\|} \qquad (4.8)$$

with $k_z \ge b_d$ and $\mathbf{r}(t)$ defined as the instantaneous performance measure (2.22). Let the adaptive learning rule for neural network weights be given by

$$\dot{\hat{W}} = \mathbf{F}\varphi(\mathbf{x})\mathbf{B}^T \mathbf{P}(\mathbf{q})\tilde{\mathbf{z}} - \kappa\|\tilde{\mathbf{z}}\|\hat{W} \qquad (4.9)$$

with $\mathbf{F} = \mathbf{F}^T > 0$ a design vector determining the learning rate and $\kappa > 0$ a design parameter governing the speed of convergence. Then the errors $\mathbf{e}(t)$, $\mathbf{r}(t)$, and $\tilde{W}(t)$ are *Uniformly Ultimately Bounded*. Moreover, the errors $\mathbf{e}(t)$, $\mathbf{r}(t)$ can be made arbitrarily small by adjusting the performance index weighting matrices $\mathbf{Q}$ and $\mathbf{R}$ properly.

Proof: Consider the following positive definite Lyapunov function

$$L(\tilde{\mathbf{z}},\tilde{W},t) = \tfrac{1}{2}\tilde{\mathbf{z}}^T \begin{bmatrix} \mathbf{K}_{n\times n} & \mathbf{0}_{n\times n} \\ \mathbf{0}_{n\times n} & \mathbf{M(q)} \end{bmatrix} \tilde{\mathbf{z}} + \tfrac{1}{2}tr(\tilde{W}^T \mathbf{F}^{-1}\tilde{W}) \qquad (4.10)$$

where $\mathbf{K}$ is positive definite and symmetric given by (3.14). The time derivative $\dot{L}$ of Lyapunov function becomes

$$\dot{L} = \tilde{\mathbf{z}}^T \mathbf{P(q)}\dot{\tilde{\mathbf{z}}} + \tfrac{1}{2}\tilde{\mathbf{z}}^T \dot{\mathbf{P}}\mathbf{(q)}\tilde{\mathbf{z}} + tr(\tilde{W}^T \mathbf{F}^{-1}\dot{\tilde{W}}). \qquad (4.11)$$

Evaluating (4.11) along the trajectory of (4.7) yields

$$\dot{L} = \tilde{\mathbf{z}}^T \mathbf{P(q)}\mathbf{A}\tilde{\mathbf{z}} - \tilde{\mathbf{z}}^T \mathbf{P(q)}\mathbf{B}\mathbf{R}^{-1}\mathbf{B}^T \mathbf{P}\tilde{\mathbf{z}} + \tfrac{1}{2}\tilde{\mathbf{z}}^T \dot{\mathbf{P}}\mathbf{(q)}\tilde{\mathbf{z}}$$
$$+ \tilde{\mathbf{z}}^T \mathbf{P(q)}\mathbf{B}\left[\tilde{W}^T \varphi(\mathbf{x}) + \varepsilon(\mathbf{x}) + \tau_{dis} + \nu(t)\right] + tr(\tilde{W}^T \mathbf{F}^{-1}\dot{\tilde{W}}). \qquad (4.12)$$

Using $\tilde{\mathbf{z}}^T \mathbf{P(q)}\mathbf{A}\tilde{\mathbf{z}} = \tfrac{1}{2}\tilde{\mathbf{z}}^T \{\mathbf{A}^T \mathbf{P(q)} + \mathbf{P(q)}\mathbf{A}\}\tilde{\mathbf{z}}$ and from the Riccati equation (3.11), we have

$$\tfrac{1}{2}\mathbf{A}^T \mathbf{P(q)} + \tfrac{1}{2}\mathbf{P(q)}\mathbf{A} + \tfrac{1}{2}\dot{\mathbf{P}}\mathbf{(q)} = -\tfrac{1}{2}\mathbf{Q} + \tfrac{1}{2}\mathbf{P(q)}\mathbf{B}\mathbf{R}^{-1}\mathbf{B}^T \mathbf{P(q)}. \qquad (4.13)$$

Then the time derivative of Lyapunov function becomes

$$\dot{L} = -\tfrac{1}{2}\tilde{\mathbf{z}}^T \mathbf{Q}\tilde{\mathbf{z}} - \tfrac{1}{2}\tilde{\mathbf{z}}^T \mathbf{P(q)}\mathbf{B}\mathbf{R}^{-1}\mathbf{B}^T \mathbf{P(q)}\tilde{\mathbf{z}} + \tilde{\mathbf{z}}^T \mathbf{P(q)}\mathbf{B}\{\varepsilon(\mathbf{x}) + \tau_{dis} + \nu(t)\}$$
$$+ tr\{\tilde{W}^T (\mathbf{F}^{-1}\dot{\tilde{W}} + \varphi(\mathbf{x})\mathbf{B}^T \mathbf{P(q)}\tilde{\mathbf{z}}\}. \qquad (4.14)$$

In the above derivation, the following trace property is used

$$trace(\mathbf{D}^T \mathbf{a}\mathbf{b}^T) = \mathbf{a}^T \mathbf{D}\mathbf{b} \qquad (4.15)$$

where $\mathbf{a} \in \mathfrak{R}^n$, $\mathbf{b} \in \mathfrak{R}^\ell$, and $\mathbf{D} \in \mathfrak{R}^{n\times\ell}$. Applying the robustifying term (4.8) and the adaptive learning rule (4.9), we obtain

$$\dot{L} \leq -\tfrac{1}{2}\|\tilde{\mathbf{z}}\|^2 \{\lambda_{min}(\mathbf{Q}) + \lambda_{min}(\mathbf{R}^{-1})\} + \|\tilde{\mathbf{z}}\|\varepsilon_{max} + \kappa\|\tilde{\mathbf{z}}\|\left(\|\tilde{W}\|_F W_M - \|\tilde{W}\|_F^2\right). \qquad (4.16)$$

The following inequality is used in the above derivation:

$$tr(\tilde{W}^T (W - \tilde{W})) = <\tilde{W}, W>_F -\|\tilde{W}\|_F^2 \leq \|\tilde{W}\|_F W_M - \|\tilde{W}\|_F^2. \qquad (4.17)$$

Completing the square terms yields

$$\dot{L} \leq -\tfrac{1}{2}\|\tilde{\mathbf{z}}\|\left[\|\tilde{\mathbf{z}}\|\{\lambda_{min}(\mathbf{Q}) + \lambda_{min}(\mathbf{R}^{-1})\} + \kappa(\|\tilde{W}\|_F - \tfrac{1}{2}W_M)^2 - \varepsilon_{max} - \tfrac{1}{4}\kappa W_M^2\right] \qquad (4.18)$$

which is guaranteed negative as long as either (4.19) or (4.20) holds

$$\|\tilde{\mathbf{z}}\| \geq (\varepsilon_{max} + \tfrac{1}{4}\kappa W_M^2)/\{\lambda_{min}(\mathbf{Q}) + \lambda_{min}(\mathbf{R}^{-1})\} \equiv B_{\tilde{\mathbf{z}}} \qquad (4.19)$$

$$\|\tilde{W}\|_F \geq \sqrt{\varepsilon_{max} + \tfrac{1}{4}\kappa W_M^2} + \tfrac{1}{2}W_M \equiv B_{\tilde{W}} \qquad (4.20)$$

where or $B_{\tilde{\mathbf{z}}}$ and $B_{\tilde{W}}$ are convergence regions of the performance measure and the weight estimation error, respectively. According to a standard Lyapunov theory extension [14, 21], this demonstrates *Uniformly Ultimate Boundedness* of $\mathbf{e}(t)$, $\mathbf{r}(t)$ and $\tilde{W}(t)$. ∎

Remarks: Although the stability analysis is similar to that of standard adaptive technique, there are fundamental differences which render the proposed CMAC neural controller universal and reusable.

1) The proposed CMAC neural controller does not require the linearity in unknown system parameters which is required in the standard adaptive techniques. The unknown nonlinearity appearing in the system is assumed to be linear with respect to basis functions in neural networks. However, the functional approximation capability holds for all continuous functions over a compact set. Hence the neural control techniques here can be applied to systems with nonlinearities which may not be linearly parameterizable.

2) The persistent excitation of learning systems has been a challenging issue since any learning algorithms for nonlinear plant have a tendency to stagnate in a "preferred" region. In other words, the more a controller learns, the more it tends to drive the system along a specific trajectory in the state space. As a result, the controller loses the control knowledge of other regions. Our adaptive learning rule (4.9) is in a form of *e*-modification where the persistency of excitation condition is not needed for adaptive control.

3) The OCTC is globally asymptotically stable if $h(\mathbf{x})$ is fully known, whereas the neural adaptive controller is *UUB*. In both cases there is a convergence of tracking errors. *UUB* is a notion of stability in the practical sense that is usually sufficient for the performance of closed-loop systems, provided that the bound on system states is small enough.

4) The external control torque has nonlinear term and linear term that can be calculated with the Riccati equation (3.11). The matrices Λ are easily computed from the weighting matrices of control optimization. The closed-loop properties may be effectively chosen with the weighting matrices, $\mathbf{Q}$ and $\mathbf{R}$. In turn, these matrices may be chosen according to general design experience in the linear quadratic optimal control theory [14].

5) Robotic manipulators are subjected to structured and/or unstructured uncertainties in all applications. Structured uncertainty is defined as the case of a correct dynamical model but with parameter uncertainty due to tolerance variations in the manipulator link properties, unknown loads, and so on. Unstructured uncertainty describes the case of unmodeled dynamics which result from the presence of high-frequency modes in the manipulator, nonlinear friction. The adaptive optimizing feature of the proposed neural controller is suitable even without full knowledge of the system dynamics.

6) From Barron's results [3] there exist lower bounds of order $(1/N_A)^{2/n}$ on the approximation error ε_{max} if only the parameters of a linear combination of basis functions are adjusted. This is not a limitation for estimation and control purposes. Our stability proof shows that the effect of the bounds on the approximation error can be alleviated by the judicious choice of weighting matrices $\mathbf{Q}$ and $\mathbf{R}$. (see (4.19) where ε_{max} is divided by $\lambda_{min}(\mathbf{Q}) + \lambda_{min}(\mathbf{R}^{-1})$).

7) The proposed adaptive learning scheme (4.9) does not require any supervisory teaching signal or known ideal output for the CMAC neural controller. Instead, the

instantaneous performance measure $\mathbf{r}(t)$ is backpropagated through the neural controller as shown in the first term in (4.9). Therefore the CMAC controller does not copy the function of the linear optimal controller. The gradient information [21] on the unknown robotic system, which is susceptible to external disturbances or unknown high frequency components is not needed. Note that Kawato *et al.* [9] used the feedback control law $\mathbf{u}(t)$ to train the neural controller in a preliminary off-line learning phase.

8) It is emphasized that the neural weight values may be initialized at zero, and stability will be maintained by the optimal controller $\mathbf{u}*(t)$ in the performance measurement loop until the neural network learns. This means that there is no off-line learning or trial and error phase, which often requires a long time in other works.

9) The advantage of the CMAC control scheme over other existing neural network architectures is that the number of adjustable parameters, i.e. weight values, is significantly less since only weights in the output layer are to be adjusted. The CMAC approach provides a reasonable compromise between complexity of learning and efficiency of function approximation. The learning algorithm can be simple since the functional form for CMAC is linear in the parameters. It is very suitable for closed-loop control.

10) It is emphasized that the proposed neural control scheme can be easily implemented in a parallel distributed processing machine, since the CMAC computation and the optimal control law computation are essentially in parallel. Furthermore, the control scheme does not require the enormous memory size of the table look-up method.

5 Simulation Results

In this section, we show some simulation results of the proposed control scheme using a 2-link manipulator. The objective is to follow the given reference trajectory. The manipulator, shown in Figure 4, is modeled as two rigid links with masses concentrated at their ends with masses m_1, m_2 (kg), lengths ℓ_1, ℓ_2 (m), angular positions q_1, q_2 (rad), and torques τ_1, τ_2 (N·m). Simulation programs are written in Turbo C using a fourth-order, fixed step-size Runge-Kutta integrator with a time step of 0.005 second and run on IBM-PC compatible machine.

The cost functional to be minimized is assumed to be the following:

$$J(\mathbf{u}) = \tfrac{1}{2} \int_{t_0}^{\infty} (\tilde{\mathbf{z}}^T \mathbf{Q} \tilde{\mathbf{z}} + \mathbf{u}^T \mathbf{R} \mathbf{u}) dt \qquad (5.1)$$

with $\mathbf{Q}$ and $\mathbf{R}$ specified in Theorem 1. The dynamic equations for an n-link manipulator are [13]

$$\mathbf{M}(\mathbf{q})\ddot{\mathbf{q}} + \mathbf{V}_m(\mathbf{q}, \dot{\mathbf{q}})\dot{\mathbf{q}} + \mathbf{F}(\mathbf{q}, \dot{\mathbf{q}}) + \mathbf{g}(\mathbf{q}) + \tau_{dis}(t) = \tau(t) \qquad (5.2)$$

where the dynamical matrices for 2-link robot are

$$\mathbf{M(q)} = \begin{bmatrix} (m_1 + m_2)\ell_1^2 + m_2\ell_2^2 + 2m_2\ell_1\ell_2 c_2 & m_2\ell_2^2 + m_2\ell_1\ell_2 c_2 \\ m_2\ell_2^2 + m_2\ell_1\ell_2 c_2 & m_2\ell_2^2 \end{bmatrix},$$

$$\mathbf{V_m(q,\dot{q})} = \begin{bmatrix} -\dot{q}_2 m_2\ell_1\ell_2 s_2 & -(\dot{q}_1 + \dot{q}_2)m_2\ell_1\ell_2 s_2 \\ \dot{q}_1 m_2\ell_1\ell_2 s_2 & 0 \end{bmatrix},$$

$$\mathbf{g(q)} = \begin{bmatrix} (m_1 + m_2)g\ell_1 c_1 + m_2 g\ell_2 c_{12} \\ m_2 g\ell_2 c_{12} \end{bmatrix} \qquad (5.3)$$

and $c_2 \equiv cos(q_2)$, $s_2 \equiv sin(q_2)$, and $c_{12} \equiv cos(q_1 + q_2)$. The selected parameters for mass and length are $m_1 = 1$ (kg), $m_2 = 2.3$ (kg), $\ell_1 = 1$ (m) and $\ell_2 = 1$ (m).

In order to study the robustness of the proposed controller, an external disturbance and frictions are added to the system dynamics as follows:

$$\tau_{dis}(t) = [8sin(2t) \quad 8cos(t)]^T \qquad (5.4)$$

$$\mathbf{F(q,\dot{q})} = diag[2 \quad 2]*\dot{q} + 1.5*sgn(\dot{q}) \qquad (5.5)$$

where $sgn(x)$ is a signum function defined by

$$sgn(x) = \begin{cases} +1 & if \ x \geq 0 \\ -1 & otherwise \end{cases}. \qquad (5.6)$$

We pick up the weighting matrices $\mathbf{Q}$, $\mathbf{R}$ as follows;

$$\mathbf{Q}_{11} = \begin{bmatrix} 10 & 2 \\ 2 & 10 \end{bmatrix}, \ \mathbf{Q}_{12} = \begin{bmatrix} -4 & 4 \\ 3 & -6 \end{bmatrix}, \ \mathbf{Q}_{21} = \mathbf{Q}_{12}^T, \ \mathbf{Q}_{22} = \mathbf{R}^{-1} = \begin{bmatrix} 30 & 0 \\ 0 & 30 \end{bmatrix}. \qquad (5.7)$$

Solving the matrices $\mathbf{K}$ and Λ using MatLab [32] yields

$$\mathbf{K} = \begin{bmatrix} 4 & -3.5 \\ -3.5 & 6 \end{bmatrix} \ and \ \Lambda = \begin{bmatrix} 2.9106 & 1.8979 \\ 1.8979 & 1.9404 \end{bmatrix}. \qquad (5.8)$$

Theorems 1 and 2 are valid for this example so that stable optimal control can be anticipated. For all the following simulations, the identified gains (5.7), (5.8) of the optimal feedback control law were used.

The motion problem considered is for the robot end-effector to track a point on a circle centered at $x = y = 1.0 \, m$ and radius $0.05 \, m$ which turns $\frac{1}{2}$ times per second in slow motion and 2 times per second in fast motion. It was pointed out that control system performance may be quite different in low speed and high speed motion. Therefore, we carry out simulations for two circular trajectories.

The desired low speed position profiles in Cartesian coordinates are

$$x_{des}(t) = 1.0 + 0.05sin(\pi t) \qquad y_{des}(t) = 1.0 + 0.05cos(\pi t) \qquad (5.9)$$

and the high speed position profiles are

$$x_{des}(t) = 1.0 + 0.05sin(4\pi t) \qquad y_{des}(t) = 1.0 + 0.05cos(4\pi t). \qquad (5.10)$$

By solving the inverse kinematics [13], we obtain the desired joint angle trajectory in fast motion as shown in Figure 4 (b) and (c).

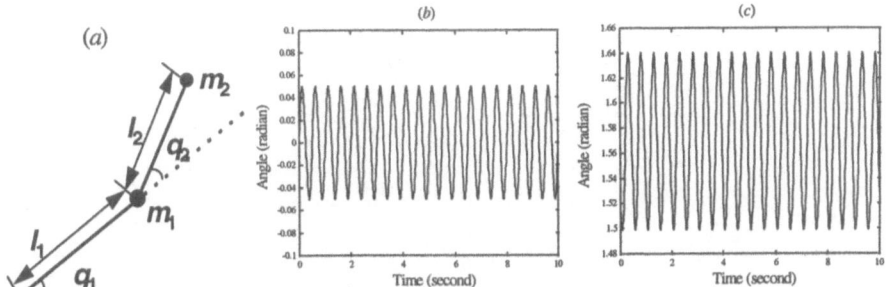

Figure 4. (a) two-link planar elbow arm (b) desired position trajectory for joint #1 (fast motion) (c) desired position trajectory for joint #2 (fast motion).

The responses of the OCTC, where all nonlinearities are exactly known, are shown in Figure 5 without external disturbances and friction forces. Only tracking errors are shown due to space limitation. The simulation was performed in low speed and high speed. As expected from the stability analysis, after a transient due to error in initial conditions, the position errors tend asymptotically to zero.

To show the effect of unstructured uncertainties, we dropped a term, $(m_1 + m_2)g\ell_1 c_1$ in gravity forces. The simulation results are shown in Figure 6 (a) in low speed. Note that the accurate modeling is necessary to keep a good performance with OCTC. Figure 6 (b) shows the effect of external disturbances and friction forces which is difficult to model and compensate. This is corrected by adding a CMAC neural network as follows.

Next the adaptive self-optimizing capability of the CMAC neural network controller was simulated. Basically, the CMAC architecture can be characterized by

- Number of input space : $\mathbf{x} = [\mathbf{q}^T \quad \dot{\mathbf{q}}^T]$
- Number of partitions for each input space : $N_i = 3$, $i = 1, \cdots, 4$
- Number of association point : $N_A = 3 \times 3 \times 3 \times 3$
- Receptive field basis functions : $\mu_{i,j}(x_i) = exp\left[-\dfrac{(x_i - m_{i,j})^2}{\sigma_{i,j}^2} \right]$ with $\sigma_{i,1} = 2$,

  $\sigma_{i,2} = 2$, $\sigma_{i,3} = 2$ and $m_{i,1} = -1$, $m_{i,2} = 0$, $m_{i,3} = 1$, $i = 1, \cdots, 4$.
- Learning rate in the weight tuning law: $\mathbf{F} = diag[8000 \quad 8000]$ and $\kappa = 0.0001$
- Simulation time: 20 seconds.

The results in Figure 7 and Figure 8, clearly show the capability of the CMAC neural network controller for overcoming uncertainties, both structured and unstructured. Note that the problem noted in Figure 6 with OCTC does not arise here as all the nonlinearities are assumed unknown to the CMAC neural controller.

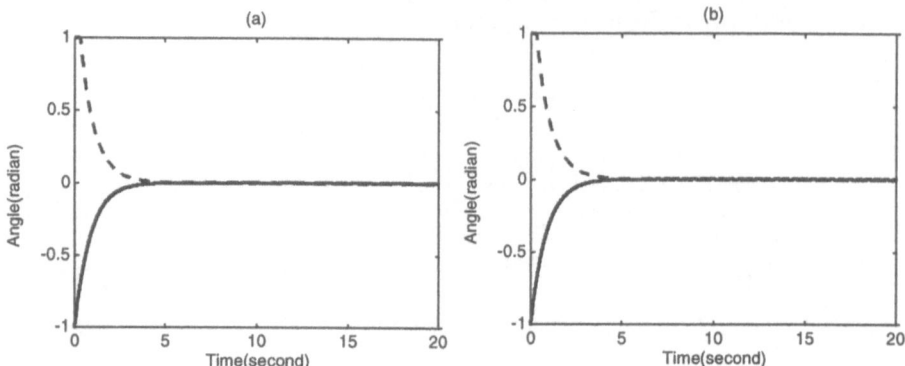

Figure 5. Performance of OCTC (3.23) (a) tracking error for slow motion
(b) tracking error for fast motion (solid : joint 1, dotted : joint 2).

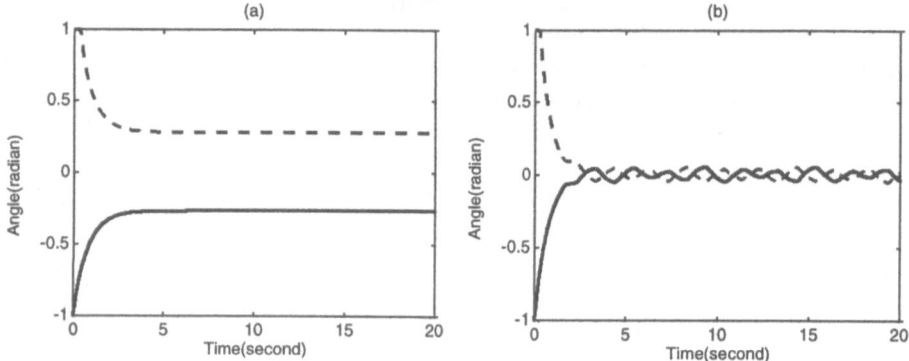

Figure 6. Performance of OCTC (3.23) (a) tracking error with modeling error for
slow motion (b) tracking error with disturbance and friction for slow motion
(solid : joint 1, dotted : joint 2).

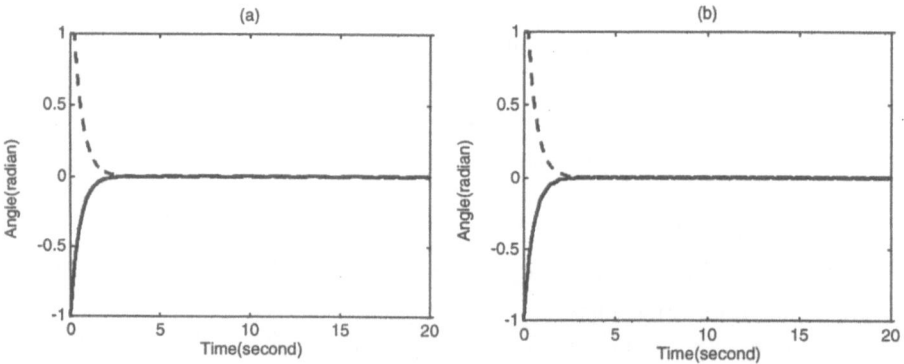

Figure 7. Performance of CMAC neural network controller (4.4) (a) tracking error
for slow motion (b) tracking error for fast motion (solid : joint 1, dotted : joint 2).

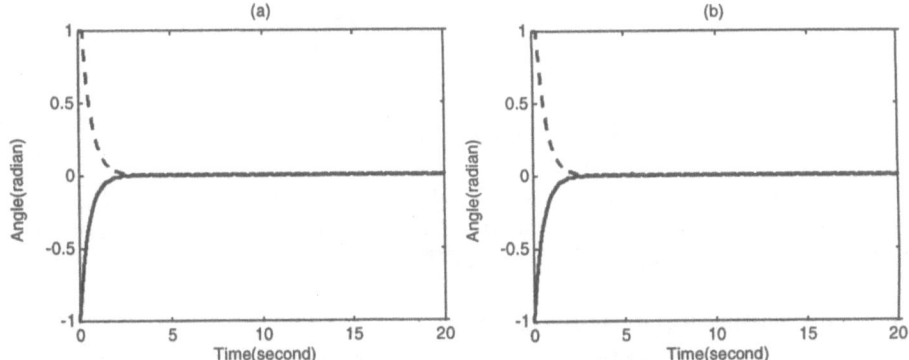

Figure 8. Performance of CMAC neural network controller (4.4) (a) tracking error with disturbance and friction for fast motion (b) tracking error of mass variation (m_2;$2.3 \rightarrow 4.0 Kg$, at 5 second, m_2;$4.0 \rightarrow 2.3 Kg$, at 12 second) with disturbance and friction for fast motion (solid : joint 1, dotted : joint 2).

6   Conclusion

We have developed an hierarchical intelligent control scheme for a robotic manipulator using H-J-B optimization process and the CMAC neural network. It has been shown that the entire closed-loop system behavior depends on the user specified performance index $\mathbf{Q}$ and $\mathbf{R}$ through the critic gain matrix Λ. The Lyapunov function for the stability of the overall system is automatically generated by the weighting matrices $\mathbf{Q}$ and $\mathbf{R}$. In the derivation of the optimal computed torque controller, it has been assumed that nonlinearities in the robotic manipulator are completely known. However even with the knowledge about the nonlinearities, it is difficult to achieve the control objective in the presence of modeling uncertainties and frictional forces. The salient feature of the CMAC neural H-J-B design is that the control objective is obtained with completely unknown nonlinearities in the robotic manipulator. The proposed neural adaptive learning shows both robustness and adaptation to changing system dynamics. To that end, a critic signal is incorporated into the adaptive learning scheme. The application potential of the proposed methodology lies in the control design in areas such as robotics and flight control and in motion control analysis (e.g., of biomechanics).

APPPENDIX I

PROOF OF LEMMA I

The approach taken here will be to use the Hamilton-Jacobi-Bellman (H-J-B) equation of dynamic programming to show that if a matrix **P** satisfies the hypothesis of *Lemma 1*, then (3.12) gives the optimal control law [6, 8, 17].

The theorem claims that the H-J-B equation

$$-\frac{\partial V(\tilde{\mathbf{z}},t)}{\partial t} = \min_{\mathbf{u}}[L(\tilde{\mathbf{z}},\mathbf{u}) + \frac{\partial V(\tilde{\mathbf{z}},t)}{\partial \tilde{\mathbf{z}}}\dot{\tilde{\mathbf{z}}}]$$
(A1.1)

is satisfied for a function

$$V = \tfrac{1}{2}\tilde{\mathbf{z}}^T \mathbf{P}(\mathbf{q})\tilde{\mathbf{z}} = \tfrac{1}{2}\tilde{\mathbf{z}}^T \begin{bmatrix} \mathbf{K} & \mathbf{0} \\ \mathbf{0} & \mathbf{M}(\mathbf{q}) \end{bmatrix}\tilde{\mathbf{z}}$$
(A1.2)

where

$$\min_{\mathbf{u}}[L(\tilde{\mathbf{z}},\mathbf{u}) + \frac{\partial V(\tilde{\mathbf{z}},t)}{\partial \tilde{\mathbf{z}}}\dot{\tilde{\mathbf{z}}}] = H(\tilde{\mathbf{z}},\mathbf{u}^*,\frac{\partial V(\tilde{\mathbf{z}},t)}{\partial \tilde{\mathbf{z}}},t).$$
(A1.3)

To derive optimal control law, the partial derivatives of the function V need to be evaluated. Here we have the time derivative of the function V

$$\frac{dV}{dt} = \frac{\partial V}{\partial t} + \frac{\partial V}{\partial \tilde{\mathbf{z}}}\dot{\tilde{\mathbf{z}}}.$$
(A1.4)

The gradient of V with respect to the error state $\tilde{\mathbf{z}}$ is

$$\frac{\partial V(\tilde{\mathbf{z}},t)}{\partial \tilde{\mathbf{z}}} = \tilde{\mathbf{z}}^T \mathbf{P}(\mathbf{q}) + \tfrac{1}{2}\tilde{\mathbf{z}}^T \mathbf{D}$$
(A1.5)

with

$$\mathbf{D} = \begin{bmatrix} \dfrac{\partial \mathbf{P}(\mathbf{q})}{\partial e_1}\tilde{\mathbf{z}} & \cdots & \dfrac{\partial \mathbf{P}(\mathbf{q})}{\partial e_n}\tilde{\mathbf{z}} & \mathbf{0} & \cdots & \mathbf{0} \end{bmatrix} = \begin{bmatrix} \mathbf{D}_1 & \mathbf{0} \end{bmatrix}.$$
(A1.6)

In (A1.6) **D** has dimension $2n \times 2n$, **0** is a $2n \times 1$ zero vector and the notation $\partial \mathbf{P}(\mathbf{q})/\partial e_1$ is used to represent the $2n \times 2n$ matrix whose elements are partial derivatives of the elements of **P** w.r.t. e_i.

A candidate for the Hamiltonian H (3.7) is the sum of (A1.5) and the Lagrangian (3.5). Now we are ready to evaluate how H depends on $\mathbf{u} \in R^n$. The $\mathbf{u} = \mathbf{u}^*$ for which H has its minimum values is obtained from the partial derivative w.r.t. $\mathbf{u}$.

Since $\mathbf{u}$ is unconstrained (A1.3) requires that

$$\frac{\partial H}{\partial \mathbf{u}}(\tilde{\mathbf{z}},\mathbf{u}^*,\frac{\partial V(\tilde{\mathbf{z}},t)}{\partial \tilde{\mathbf{z}}},t) = 0$$
(A1.7)

which gives a candidate for the optimal control

$$\mathbf{u}*(t) = -\mathbf{R}^{-1}\mathbf{B}^T \frac{\partial V(\tilde{\mathbf{z}},t)^T}{\partial \tilde{\mathbf{z}}} . \tag{A1.8}$$

Since

$$\frac{\partial^2 H}{\partial \mathbf{u}^2} = \mathbf{R} > 0 . \tag{A1.9}$$

We know that (A1.3) is satisfied by $\mathbf{u}*(t)$ given (A1.8). Substituting (A1.5) and (A1.6) into (A1.8) gives

$$\mathbf{u}* = -\mathbf{R}^{-1}\mathbf{B}^T \mathbf{P}(\mathbf{q})\tilde{\mathbf{z}} . \tag{A1.10}$$

Notice that the relation

$$\mathbf{DB} = \mathbf{D}_1 *0 + 0*\mathbf{M}(\mathbf{q}) = 0 \tag{A1.11}$$

is used.

A necessary and sufficient condition for optimality is that the chosen value function V satisfies (3.6). Substituting (3.7) into (3.6) yields

$$\frac{\partial V(\tilde{\mathbf{z}},t)}{\partial t} + \frac{\partial V(\tilde{\mathbf{z}},t)}{\partial \tilde{\mathbf{z}}}\dot{\tilde{\mathbf{z}}} + L(\tilde{\mathbf{z}},\mathbf{u}*) = 0 \tag{A1.12}$$

where it is understood that the partial derivatives of V in (A1.12) are being evaluated along the optimal control $\mathbf{u}*$. Substituting (A1.4) into (A1.12), we obtain

$$\tilde{\mathbf{z}}^T \mathbf{P}(\mathbf{q})\dot{\tilde{\mathbf{z}}} + \frac{1}{2}\tilde{\mathbf{z}}^T \dot{\mathbf{P}}(\mathbf{q})\tilde{\mathbf{z}} + L(\tilde{\mathbf{z}},\mathbf{u}*) = 0 . \tag{A1.13}$$

Inserting (3.3), (3.5) and (A1.10) into (A1.13) gives

$$\tilde{\mathbf{z}}^T \mathbf{P}(\mathbf{q})\mathbf{A}\tilde{\mathbf{z}} + \frac{1}{2}\tilde{\mathbf{z}}^T \{\dot{\mathbf{P}}(\mathbf{q}) + \mathbf{Q} - \mathbf{P}(\mathbf{q})\mathbf{B}\mathbf{R}^{-1}\mathbf{B}^T \mathbf{P}(\mathbf{q})\}\tilde{\mathbf{z}} = 0 . \tag{A1.14}$$

Since $\tilde{\mathbf{z}}^T \mathbf{P}(\mathbf{q})\mathbf{A}\tilde{\mathbf{z}} = \frac{1}{2}\tilde{\mathbf{z}}^T \{\mathbf{A}^T \mathbf{P}(\mathbf{q}) + \mathbf{P}(\mathbf{q})\mathbf{A}\}\tilde{\mathbf{z}}$, (A1.14) can be written as

$$\frac{1}{2}\tilde{\mathbf{z}}^T \{\dot{\mathbf{P}}(\mathbf{q}) + \mathbf{A}^T \mathbf{P}(\mathbf{q}) + \mathbf{P}(\mathbf{q})\mathbf{A} + \mathbf{Q} - \mathbf{P}(\mathbf{q})\mathbf{B}\mathbf{R}^{-1}\mathbf{B}^T \mathbf{P}(\mathbf{q})\}\tilde{\mathbf{z}} = 0 . \tag{A1.15}$$

We can summarize by stating that if a matrix $\mathbf{P}$ can be found which satisfies (A1.15) $\forall t \in (t_0, \infty)$, then the value function given in (A1.2) satisfies the H-J-B equation (A1.1). In this case the desired optimal control is given by (A1.10). Note that if the matrix $\mathbf{P}$ satisfies the algebraic Riccati equation (3.11), then $\mathbf{P}$ satisfies (A1.15). This completes the proof. ∎

APPPENDIX II

PROOF OF THEOREM I

From *Lemma 1* it is known that

$$V = \tfrac{1}{2}\tilde{\mathbf{z}}^T \mathbf{P}(\mathbf{q})\tilde{\mathbf{z}} = \tfrac{1}{2}\tilde{\mathbf{z}}^T \begin{bmatrix} \mathbf{K} & 0 \\ 0 & \mathbf{M}(\mathbf{q}) \end{bmatrix} \tilde{\mathbf{z}} \qquad (A2.1)$$

solves the H-J-B equation for $\mathbf{K} = \mathbf{K}^T$, Λ solving the matrix equation from the quadratic form

$$\tilde{\mathbf{z}}^T \left(\mathbf{P}(\mathbf{q})\mathbf{A} + \mathbf{A}^T\mathbf{P}(\mathbf{q})^T - \mathbf{P}(\mathbf{q})\mathbf{B}\mathbf{R}^{-1}\mathbf{B}^T\mathbf{P}(\mathbf{q}) + \dot{\mathbf{P}}(\mathbf{q}) + \mathbf{Q} \right)\tilde{\mathbf{z}} = 0 . \qquad (A2.2)$$

The optimal feedback control law that minimizes $J(\mathbf{u})$ is

$$\mathbf{u}*(t) = -\mathbf{R}^{-1}\mathbf{B}^T\mathbf{P}(\mathbf{q})\tilde{\mathbf{z}}(t) . \qquad (A2.3)$$

Let the weighting matrices $\mathbf{Q}$, $\mathbf{R}$ be given by (3.19).

Substitution of expressions for matrices $\mathbf{A}$, $\mathbf{B}$ (3.3) and $\mathbf{P}(\mathbf{q})$ (3.10) into (A2.2), we have

$$\tilde{\mathbf{z}}^T \left(\begin{bmatrix} -\mathbf{K}\Lambda & \mathbf{K} \\ 0 & -\mathbf{V}_m(\mathbf{q},\dot{\mathbf{q}}) \end{bmatrix} + \begin{bmatrix} -\Lambda^T\mathbf{K}^T & 0 \\ \mathbf{K} & -\mathbf{V}_m^T(\mathbf{q},\dot{\mathbf{q}}) \end{bmatrix} - \begin{bmatrix} 0 & 0 \\ 0 & \mathbf{R}^{-1} \end{bmatrix} \right.$$
$$\left. + \begin{bmatrix} 0 & 0 \\ 0 & \dot{\mathbf{M}}(\mathbf{q}) \end{bmatrix} + \begin{bmatrix} \mathbf{Q}_{11} & \mathbf{Q}_{12} \\ \mathbf{Q}_{12}^T & \mathbf{Q}_{22} \end{bmatrix} \right) \tilde{\mathbf{z}} = 0 . \qquad (A2.4)$$

Whence the application of robot property 2, (2.29) shows that the matrices $\mathbf{K}$, Λ of (3.20) and (3.21) solve the algebraic Riccati equation of (A2.5)

$$\begin{bmatrix} -\mathbf{K}\Lambda & \mathbf{K} \\ 0 & 0 \end{bmatrix} + \begin{bmatrix} -\Lambda^T\mathbf{K}^T & 0 \\ \mathbf{K} & 0 \end{bmatrix} - \begin{bmatrix} 0 & 0 \\ 0 & \mathbf{R}^{-1} \end{bmatrix} = -\begin{bmatrix} \mathbf{Q}_{11} & \mathbf{Q}_{12} \\ \mathbf{Q}_{12}^T & \mathbf{Q}_{22} \end{bmatrix}. \qquad (A2.5)$$

This completes the proof. ∎

References

[1] Albus, J.S. (1975), "A new approach to manipulator control; the cerebellar model articulation controller(CMAC)," *J. Dynamic Syst., Measurement, Contr.*, Vol. 97, pp. 220-227.

[2] Barto, A.G. (1990), "Connectionist learning for control," *Neural Networks for Control*, The MIT Press, pp. 5-58.

[3] Barron, A.R. (1993), "Universal approximation bounds for superposition of a sigmoidal function," *IEEE Trans. Inform. Theory*, Vol. 39, pp. 930-945.

[4]   Chiang, C.-T. and Lin, C.-S. (1996), "CMAC with general basis functions," *Neural Networks*, vol. 9, pp. 1199-1211,.

[5]   Commuri, S., Lewis, F.L., Zhu, S.Q., and Liu, K. (1995), "CMAC neural networks for control of nonlinear dynamical systems," *Proceedings of Neural, Parallel and Scientific Computation*, Vol. 1, pp. 119-124.

[6]   Dawson, D., Grabbe, M., and Lewis, F.L. (1991), "Optimal control of a modified computed-torque controller for a robot manipulator," *Int. J. Robot. Automat.*, Vol. 6, pp. 161-165.

[7]   Hunt, K.J., Sbarbaro, D., Zbikowski, R., and Gawthrop, P.J. (1992), "Neural networks for control systems: a survey," *Automatica*, Vol. 28, pp. 1823-1836.

[8]   Johansson, R. (1990), "Quadratic optimization of motion coordination and control," *IEEE Trans. Automat. Contr.*, Vol. 35, pp. 1197-1208.

[9]   Kawato, M., Uno, Y., Isobe, M., and Suzuki, R. (1988), "Hierarchical neural network model for voluntary movement with application to robotics," *IEEE Contr. Sys. Mag.*, pp. 8-16, Apr.

[10] Kirk, D.E. (1986), *Optimal Control Theory: An Introduction*, Publisher, New York.

[11] Koditschek, D.E. (1987), *Qudaratic Lyapunov functions for mechanical systems*, Yale Univ. Tech. Rep. 703.

[12] Lane, S.H., Handelman, D.A., Gelfand, J.J. (1992), "Theory and development of higher-order CMAC neural networks," *IEEE Contr. Sys. Mag.*, pp. 23-30, Apr.

[13] Lewis, F.L., Abdallah, C.T., and Dawson, D.M. (1993), *Control of Robot Manipulators*, MacMillan, New York.

[14] Lewis, F.L. and Syrmos, V.L. (1995), *Optimal Control*, second ed., Wiley, New York.

[15] Lewis, F.L., Yesildirek, A., and Liu, K. (1996), "Multilayer neural-net robot controller with guaranteed tracking performance," *IEEE Trans. Neural Networks*, Vol. 7, pp. 388-399.

[16] Lin, C.-S. and Kim, H. (1991), "CMAC-based adaptive critic self-learning control," *IEEE Trans. Neural Networks*, Vol. 2, no. 5, pp. 530-533.

[17] Luo, G.L. and Saridis, G.N. (1985), "L-Q design of PID controllers for robot arms," *IEEE Jour. of Robot. and Automat.*, Vol. 1, no. 3, pp. 152-157.

[18] Miller, W.T., Glanz, F.H., and Kraft, L.G. (1987), "Application of a general learning algorithm to the control of robotic manipulators," *Int. J. Robot. Res..*, Vol. 6, pp. 84-98,

[19] Miller, W.T., Hewes, R.H., Glanz, F.H., and Kraft, L.G. (1990), "Real-time dynamic control of an industrial manipulator using a neural-network based learning controller," *IEEE Trans. Robot. Automat.*, Vol. 6, no. 1, pp. 1-9.

[20] Mischo, W.S. (1996), "How to adapt in neuro control a decision for CMAC," *Neural adaptive control technology*, World Scientific Publishing Co., pp. 285-314.

[21] Narendra, K.S. and Annaswamy, A.M. (1987), "A new adaptive law for robust adaptation without persistent excitation," *IEEE Trans. Automat. Contr.*, Vol. 32, no.2, pp. 134-145.

[22] Narendra, K.S. and Parthasarathy, K. (1990), "Identification and control of dynamical systems using neural networks," *IEEE Trans. Neural Networks*, Vol. 1, pp. 4-27.

[23] Newton, R. and Xu, Y. (1993), "Neural network control of a space manipulator," *IEEE Contr. Sys. Mag.*, pp. 14-22 (Dec.).

[24] Ozaki, T., Suzuki, T., Furuhashi, T., Okuma, S., and Uchikawa, Y. (1991), "Trajectory control of robotic manipulators using neural networks," *IEEE Trans. Ind. Elec.*, Vol. 38, no. 3, pp. 195-202.

[25] Polycarpou, M.M. (1996), "Stable adaptive neural control of scheme for nonlinear systems," *IEEE Trans. Automat. Contr.*, Vol. 41, pp. 447-451.

[26] Saad, M., Bigras, P., Dessaint, L.A., and Al-Haddad, K. (1994), "Adaptive robot control using neural networks," *IEEE Trans. Ind. Elec.*, Vol. 41, pp. 173-181.

[27] Sanner, R.M. and Slotine, J.-J.E. (1992), "Gaussian networks for direct adaptive control," *IEEE Trans. Neural Networks*, Vol. 3, pp. 837-863.

[28] Saridis, G.N. and Lee, C.-S.G. (1979), "Approximation theory of optimal control for trainable manipulators," *IEEE Trans. Syst. Man Cybern.*, Vol. 9, pp. 152-159.

[29] Slotine, J.-J.E. and Li, W. (1991), *Applied Nonlinear Control.* Prentice Hall.

[30] Werbos, P.J. (1990), "Backpropagation through time: what it does and how to do it," *Proc. IEEE*, Vol. 78, no. 10, pp. 1550-1560.

[31] Wong, Y.-F. and Sideris, A. (1992), "Learning convergence in the Cerebellar Model Articulation Controller," *IEEE Trans. Neural Networks*, Vol. 3, pp. 115-121.

[32] MatLab User's Guide (1990), *Control System Toolbox*, The Mathworks, Inc., Natick, MA.

Chapter 6

HIERARCHICAL BEHAVIOR CONTROLLER IN ROBOTIC APPLICATIONS

Yasuhisa HASEGAWA
Toshio FUKUDA

Dept. of Micro System Engineering
Nagoya University
Japan

In this chapter, we propose a new controller and a learning algorithm. Our system consists of several subcontrollers to indicate the desired trajectories for robot's actuators. This algorithm selects the subcontroller which is not appropriate and needs to be tuned, by evaluating each subcontroller using multiple regression analysis based on previously obtained evaluation values. This can reduce the learning iterations by avoiding attempts to tune good subcontrollers. The proposed algorithm is applied to the problem of selecting and tuning subcontrollers at a middle layer in the hierarchical behavior controller in order to compensate imperfect initial controllers. The hierarchical behavior controller is applied to the problem of controlling a seven-link brachiation robot, which moves dynamically from branch to branch like a gibbon, a long-armed ape, swinging its body in a pendulum (Figure 1).

1 Introduction

Intelligence can be shown to grow and evolve through computational power and accumulation of knowledge to sense, decide and act in a complex and changing world. There are four elements of intelligence: sensory processing, world modeling, behavior generation and value judgment. Input to, and output from, intelligent systems are via sensors and actuators [1].

In this chapter, we focus on the behavior generating and a learning method for a behavior controller composed of multiple subcontrollers. Behavior is a range of continuous actions which are performed by a robot with multiple degrees of freedom. To control such a robot, the controller is required to have the capability for managing multiple inputs and outputs. It would be difficult work to design such a controller, because we have to consider the non-linear property among all inputs and outputs that are necessary to generate the behavior. Therefore, we set the same number of subcontrollers in the behavior controller as that of actuators which are necessary to generate that behavior.

We can consider three ways to design the subcontrollers which cooperatively make a robot to perform a desired behavior. The first is the design of behavior controller. The second is a suitable search method or optimal method; Genetic Algorithm, Backpropagation methods, Reinforcement Learning and so on, for designing the behavior controller and maximizing a given evaluation function. The third is the combination of the above two ways.

In the next section, the hierarchical behavior controller is described. In section 3, the proposed algorithm is described. In section 4, a fuzzy logic controller is described and in section 5, its learning algorithm is presented. In section 6, 7-link brachiation robot using a computer simulation is explained. Section 7 presents the design of initial controller. The proposed method is applied for obtaining a behavior controller for 7-link brachiation robot to show its effectiveness in section 8.

Figure 1. Brachiation of a Gibbon.

2   Architecture of Hierarchical Behavior Controller

Most robots have multiple degrees of freedom, which is required to perform various kinds of objective behavior or task. Therefore multiple sensors are actuators are necessary for a robot to perform these complex behaviors. The robot's controller, which deals with multiple input and output variables, also becomes complex to design and adjust according to an environment change. It is beneficial if the complex behaviors are divided into element behaviors and if behavior controllers for each element behavior are designed. In this case, a manager on the upper level is necessary to control the behavior controllers. In this way, a hierarchical structure emerges like a human society. In this section, we show a hierarchical behavior controller for complex robot systems.

The hierarchical behavior controller shown in Figure 2 consists of three layers: a planner, behavior controllers and feedback controllers. The planner in the top layer decides when and which behavior controllers in the middle layer should be used to make the robot perform the desired task. Multiple behavior controllers are on the middle layer. The behavior controller shown in Figure 3 is a behavior generator which generates a sequence of actions and it outputs the desired trajectories to actuators. The behavior controller consists of multiple subcontrollers, which have multiple input variables and an output variable, as shown in Figure 4. Each of the feedback controllers in the bottom layer makes an actuator follow the desired trajectory indicated by the behavior controller, using a linear or nonlinear feedback control method. When the desired trajectories from more than two behavior controllers are inputted to a feedback controller, the weighted mean value of them becomes the desired trajectory. The environment which is not measurable by the sensors, for example the center of gravity and the figure of the object, is calculated in the sensor fusion section and then is outputted to the planner and the behavior controller.

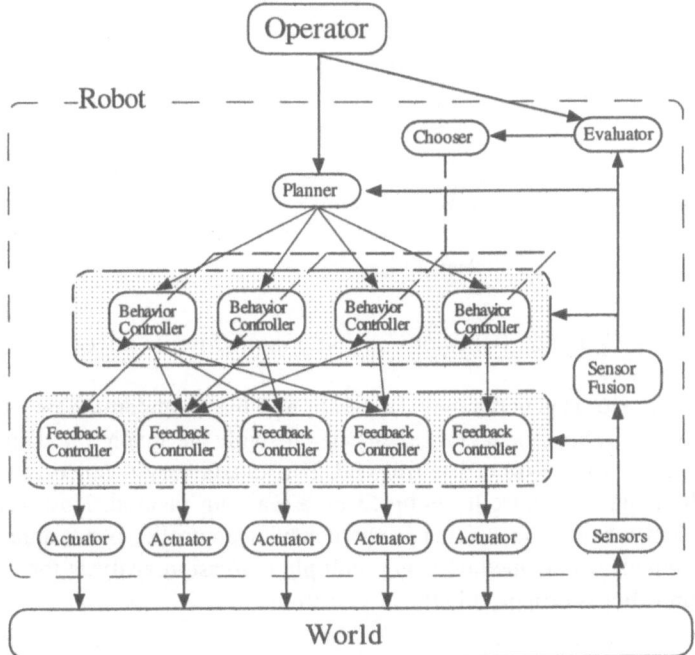

Figure 2. Architecture of Hierarchical Behavior Controller

The system behavior may be acquired by trial and error learning, but more often it is acquired from a teacher, or from written or programmed instructions. This given knowledge can reduce the searching space and enable us to obtain a desired controller in fewer learning iterations. We plan the same number of subcontrollers in the behavior controller as actuators which are necessary to generate that behavior and then design an initial controller in each subcontroller based on human knowledge. The initial controller is only to output a numerical value according to the robot states, like a mathematical function, Cerebella Model Arithmetic Computer (CMAC) [2][3], and neural networks for a fuzzy map. The fuzzy controller in the subcontroller for compensation of the initial controller is tuned by any learning method through trials. The numerical value to the feedback controller in the bottom layer is the sum of the values from the initial controller and from the fuzzy controller.

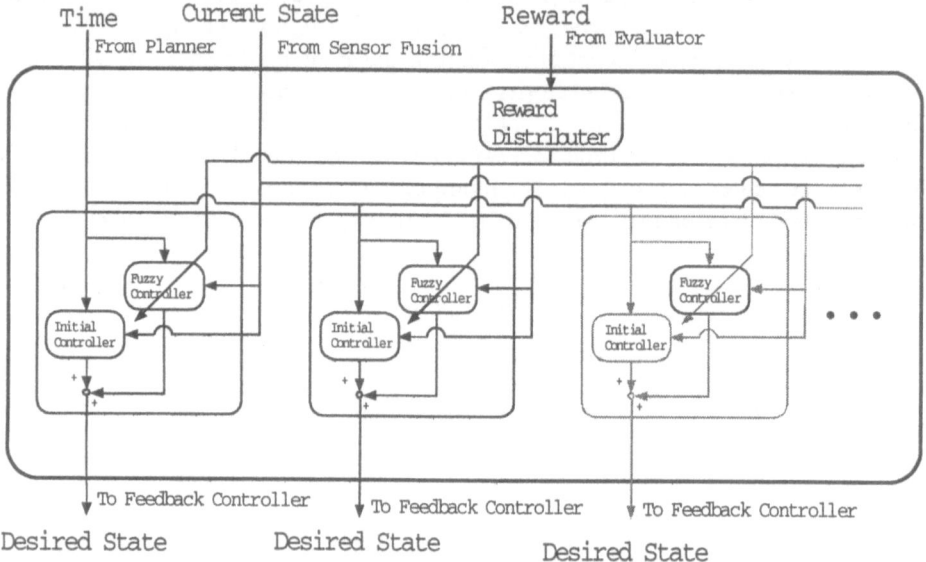

Figure 3. Behavior Controller in the Middle Layer

The system behavior is generated by cooperation of these subcontrollers. However the evaluation of the robot's behavior is carried out with the given evaluation function after the behavior of the robot has finished. If a satisfactory evaluation value is not obtained, the behavior controller is tuned by a learning method. However it is very difficult to find which subcontroller in the behavior controller is not appropriate and needs to be tuned. A new method using multiple regression analysis for the selection of the subcontroller is proposed in the next section.

3   Selection of Subcontroller

We installed some subcontrollers in the behavior controller, which generates a range of actions. Each of the subcontrollers outputs the desired trajectory of an actuator, according to multiple input variables. The desired behavior is performed by the cooperation of multiple subcontrollers in the behavior controller. If the performance of the robot is not satisfactory, the behavior controller should be tuned. However the behavior controller has multiple subcontrollers, and selection of the subcontroller that is not appropriate and that should be tuned is very difficult but important. We present an algorithm to select the inappropriate subcontroller.

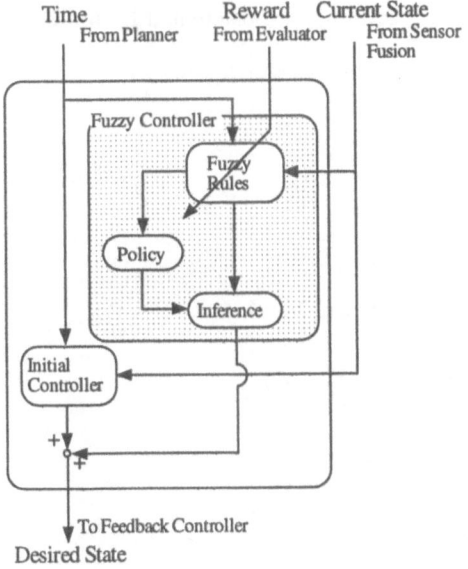

Figure 4. Sub-controller in the Behavior Controller

The first step is to evaluate each subcontroller using multiple-regression analysis based on the previously obtained evaluation values, and then choose a subcontroller to be tuned in the next trial with the probability of choice in proportion to the evaluation values of the subcontrollers. This can reduce the learning iterations by avoiding attempts to tune good subcontrollers. The method is explained below.

After tuning the subcontroller, n, we calculate the improvement in the performance of the robot. The improvement value, f_d, is

$$f_{d(x)} = f_{m(x)} - f'_{m(x)} \tag{1}$$

when f_m is the performance score calculated by the given evaluation function, f'_m is the performance evaluated in the previous trial and x is the learning time to tune the subcontroller. The evaluation value of the subcontroller is

$$f_{n(x)} = f_{n(x-1)} + f_{d(x)} \tag{2}$$

We assumed that the evaluation value, f_n, of the subcontroller is roughly proportional to its learning time, x.

$$f_{nd} = a_1 x + a_0 \tag{3}$$

The coefficient values, a_1 and a_0, are calculated by multiple-regression analysis as follows.

$$a_1 = \frac{S_{x f_m}}{S_x^2} \tag{4}$$

$$a_0 = \bar{f}_m - a_1 \bar{x} \tag{5}$$

$S_{x f_n}$ is the covariance between x and f_n.

$$S_{x f_m} = \frac{1}{N} \sum_{i=1}^{N} (x_i - \bar{x})(f_i - \bar{f}) \tag{6}$$

and S_x^2 is the variance of x.

$$S_x^2 = \frac{1}{N-1} \sum_{i=1}^{N} (x_i - \bar{x})^2 \tag{7}$$

a_1 corresponds to the prospect of improvement of the system performance when its subcontroller will be tuned in the next trial. The variance of f_m, $S_{f_m}^2$, indicates the reliability of the system improvement. The evaluation function of the subcontrollers therefore consists of two terms, a_1 and $S_{f_m}^2$ shown below:

$$S_{f_n}^2 = \frac{1}{N-1} \sum_{i=1}^{N} (f_{n(i)} - \bar{f}_{n(x)})^2 \tag{8}$$

$$P_c = a_1 - \gamma S_{f_n}^2 \tag{9}$$

where γ is a coefficient > 0.

A subcontroller tuned in the next trial is selected according to the probability, g_c in the following equation, which is based on the evaluation values, p_c.

$$g_c = \frac{P_c}{\sum_{c=1}^{C} P_c} \tag{10}$$

The flow chart of the proposed algorithm is shown in Figure 5.

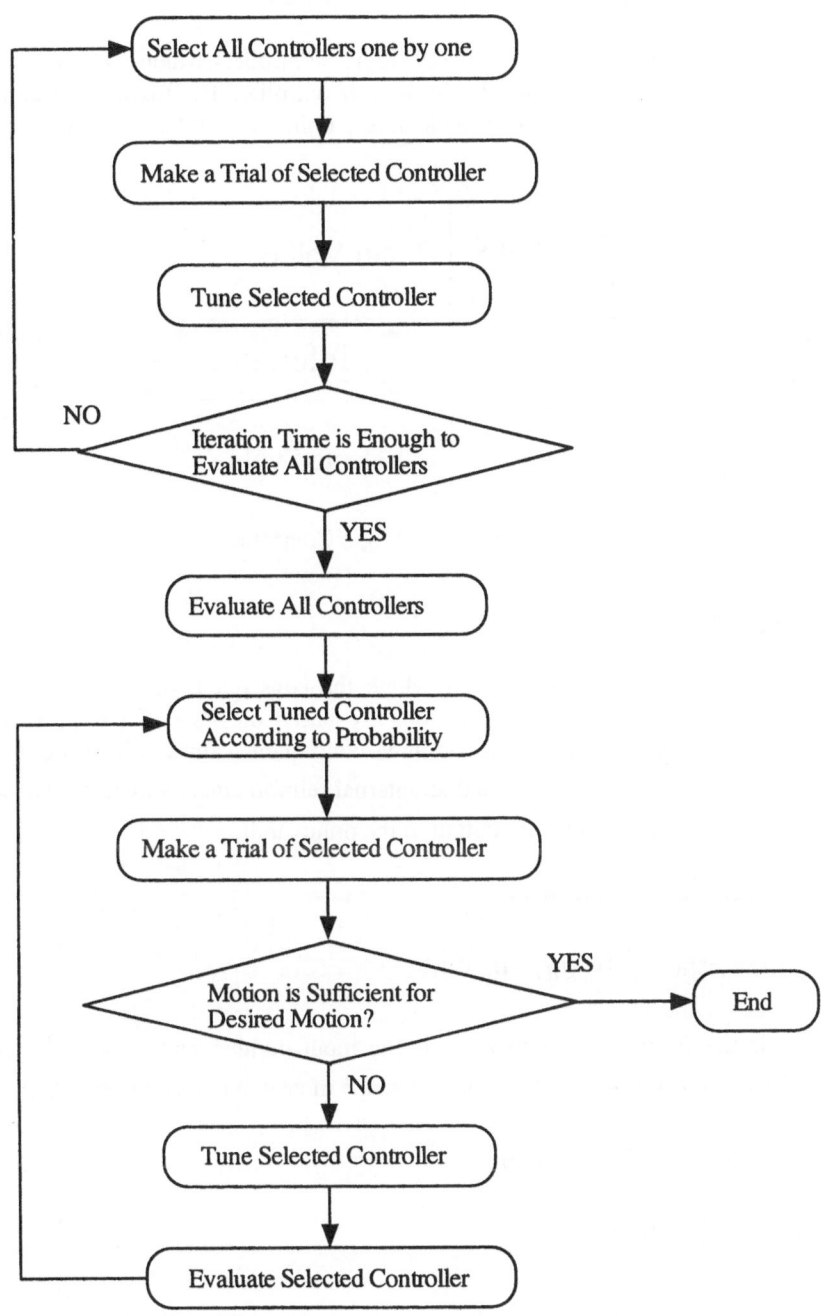

Figure 5. Flow Chart

4 Fuzzy Logic Controller

In this section, we present the fuzzy logic controller which is used for the compensation of the initial controller in the subcontroller. The fuzzy logic controller is composed of three units: Fuzzy Rule Base, Policy and Inference unit shown in Figure 6.

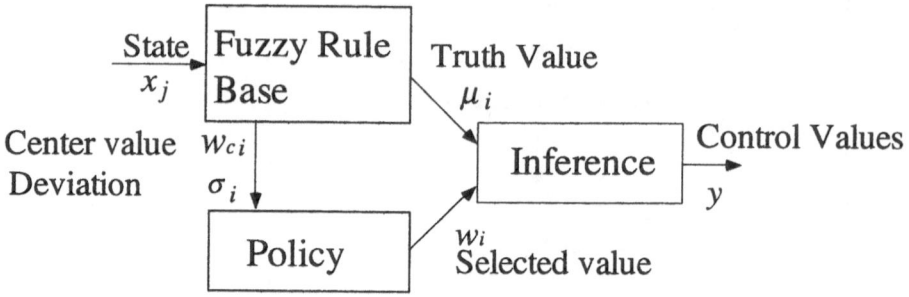

Figure 6. Fuzzy Logic Controller

4.1 Fuzzy Rule Base

We use the simplified fuzzy rules, of which the consequent parts are represented numerically and the membership functions are based on the radial basis functions (RBFs)[4]. A simplified fuzzy rule has three consequent values, which are a center value w_{ci}, a dispersion value σ_i and an internal reinforcement value, r_i. The center value and the dispersion value are output to the policy unit.

The antecedent part is expressed as

$$\mu_{ij}(t_d) = \exp\left(-b_{ij} \cdot \left(x_j(t_d) - a_{ij}\right)^2\right),$$ (11)

where i is the fuzzy rule number, j is the input number and a and b are the coefficients that determine the position and shape of each membership function.

The true value μ_i of the ith rule is

$$\mu_i(t_d) = \prod_J \mu_{ij}(t_d) \ .$$ (12)

This true value of each fuzzy rule is output to the inference unit.

4.2 Policy

In this part, the consequent values w_i are determined, and are used to infer the control values in the inference unit. In a searching mode, they are random values determined according to a probability distribution expressed by

$$g_i = \frac{1}{2\pi\sigma_i} \exp\left(-\frac{(w_i - w_{ci})^2}{2\sigma_i^2} \right). \tag{13}$$

The search area depends on the dispersion value. When the dispersion value σ_i is large, a global search about the consequent value w_i is performed. When it is small, a local search is performed. When the obtained controller controls a robot after the searching mode, the consequent values are equivalent to the center values from fuzzy rule base.

4.3 Inference

In the inference unit, control values for a robot are inferred and its output based on the true values from the fuzzy rule part and the consequent values from policy part. The control value Y of fuzzy reasoning is calculated by:

$$Y(t) = \frac{\sum_{i=1}^{I} \mu_i(t) \cdot w_i}{\sum_{i=1}^{I} \mu_i(t)}. \tag{14}$$

5   Self-Scaling Reinforcement Learning

The reinforcement learning method needs many iterations to realize the state-action map. One of its causes is the very low learning rate. If the reinforcement range of variation is known, the learning rate can be adjusted to prevent the weights from overshooting which impedes the system performance. When this is not the case, then, a very low learning rate is used, and this results in a slow learning process. When a learning algorithm attains much larger reinforcement than in the previous trials, it might be approaching the goal and should change search strategy to a local search . Therefore we propose a new algorithm to change the learning rate according to the reinforcement values obtained [5]. This self-scaling method prevents the weights from overshooting and finds the optimal solution in fewer iterations.

After a robot executes a range of objective tasks, the results are evaluated using the evaluated value f. We calculate the past performance p using eq. (15), which is the weighted mean of the past evaluation values. Furthermore we compute an inner reinforcement value r' by eq. (16) below.

$$p_{(s)} = \frac{\sum\limits_{i=0}^{s} k^{s-i} f_{(i)}}{\sum\limits_{i=0}^{s} k^{s-i}} \, , \tag{15}$$

$$r' = f_{(s)} - p_{(s-1)} \, , \tag{16}$$

where s is the trial's number and k is the positive weight <1.

The range of this reinforcement value r' is decided by the evaluation function which is arranged by the designer. It is then transformed into r in the range [-1, 1].

$$r(s) = \frac{1 - \exp\left(-\frac{r'(s)}{\beta}\right)}{1 + \exp\left(-\frac{r'(s)}{\beta}\right)} \, , \tag{17}$$

where β is a positive coefficient.

The reinforcement value is found after a robot executes a set of objective tasks or motions. Related works use fuzzy-neural networks or CMAC to compute the internal reinforcements [6][7]. Therefore we assign an internal reinforcement to every fuzzy rule in fuzzy rule base unit explained in section 4.1. The internal reinforcement, r_i, assigned to fuzzy rule, i, is updated by the following equation:

$$r_i \Leftarrow r_i (1 - \max_t \mu_i(t)) + \lambda^{(T^* - T_i)} r \max_t \mu_i(t) \tag{18}$$

where λ is a positive constant < 1, T_i is the time when the true value (eq. (12)) has the maximum value, and T^* is the time when the reinforcement is received.

The center values and the dispersion values of the consequent parts in fuzzy rules used during the robot motion are updated according to the internal reinforcement calculated by equation (18). The equations are

$$w_{ic} \Leftarrow w_{ic} + r_i (w_i - w_{ic}) \qquad \text{if } r \geq 0 \, , \tag{19}$$

$$\sigma_i \Leftarrow \sigma_i(1 - r_i) \qquad\qquad \text{if } r \ge 0. \tag{20}$$

This algorithm would converge only with above two increment equations into the local minimum around an initial position. Therefore when the reinforcement, r, is less than zero, the dispersion values of the consequent parts are increased by

$$\sigma_i \Leftarrow \sigma_i + \alpha \qquad\qquad \text{if } r < 0, \tag{21}$$

where α is positive. If the reinforcement which is much better than the average, p, is obtained, the center values of the consequent values are updated to almost the same values as those used in this trial and the dispersion values become very small. In this case, a local search is performed until the received reinforcement becomes smaller than the average. On the other hand, if the obtained reinforcement is smaller than the average, the search range is widened.

In order to validate a controller for a complicated real robot, it is necessary to perform the trials rather than computer simulations. It is useful to simulate a controller for the real robot in fewer iterations, because computer simulations can easily repeat the trials until the solution is found, but the real robot durability and time (cost) are limited. In such a case, this algorithm shows the utility of finding feasible solutions in fewer iterations, instead of finding the best solutions.

6   Seven-Link Brachiation Robot

Figure 7 shows the brachiation mobile robot (BMR) of a seven-link model used for a computer simulation in this chapter. This mobile robot can dynamically move from branch to branch like a gibbon (Figure 1), namely long-armed ape, swinging its body like a pendulum [8]-[10]. It has two arms, a body and a leg. The robot has seven degrees of freedom and six control inputs to actuators; elbows and shoulders of both arms, hip and knee. It has redundant degree of freedom to move from branch to branch. On the tips of the two arms, the grips are set to catch horizontal parallel bars. In this study, the motion of the robot is assumed to be within a vertical plane. The equation of motion is solved from Lagrange's equation of motion [11].

Table 1 covers each of the physical parameters used in the simulation, where l is the length of a link, s is the center of gravity, m is the mass and J is the moment of inertia. Table 2 shows each of the joint parameters, where D is the coefficient of viscous friction, $|\max|\tau||$ is a limit of control torque for the driving motor $\max|d\tau / dt|$ is a increase or decrease limit of control torque and $\max\theta$ and $\min\theta$ are range of the angle between a vertical line and its link.

$$\tau = \sum_{k}^{0\uparrow} \left\{ \left(J_k + m_k s_k^2 + \sum_{i}^{k\uparrow} m_i l_k^2\right) \ddot{\Theta}_k + gM_k \sin \Theta_k \right\}$$
$$+ \sum_{k}^{all} \left\{ \left(\sum_{i}^{k\downarrow} Q_{ik} + \sum_{j}^{k\uparrow} Q_{kj}\right) \ddot{\Theta}_k + \left(\sum_{i}^{k\downarrow} P_{ik} - \sum_{j}^{k\uparrow} P_{kj}\right) \dot{\Theta}_k^2 \right\} \qquad (22)$$
$$+ \tau_{friction}$$

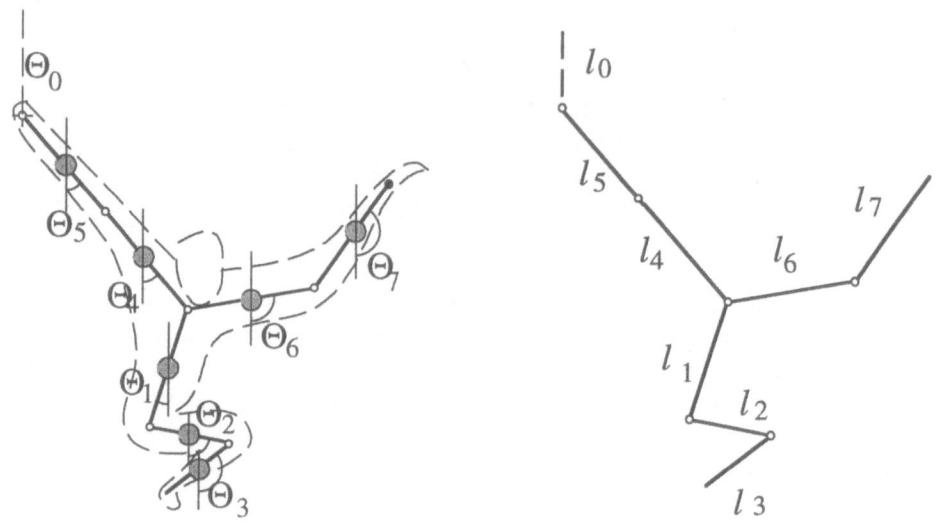

Figure 7. Seven-link Brachiation Robot

Table 1. Parameters of Seven-link Brachiation Robot

Link	1	2	3	4, 5, 6, 7
l_i [m]	0.287	0.195	0.190	0.295
s_i [m]	0.107	0.098	0.159	0.148
m_i [kg]	4.0	1.5	1.0	0.5
J_i [kg·m^2]	5	0.3	0.5	0.15

Table 2. Joint Parameters of Seven-link Brachiation Robot

Joint	05	12	23	14, 16	45, 67
D_{ij} [N·m·s/rad]	0.5	2.0	0.8	0.5	0.6
max $|\tau_{ij}|$ [N·m]	0	10	8	12	8
max $|d\tau_{ij}/dt|$ [N·m]	0	10	8	12	8
max $|\theta_{ij}|$ [deg]	-	150	0	-	150
min $|\theta_{ij}|$ [deg]	-	0	-150	-	-30

7   Design of the Initial Controller

Seven-link brachiation robot has 8 actuators involving two actuators for grips to hold or release the bar. The controller should control in cooperation with 8 actuators so that the robot can move from one branch to another branch. We designed a planner, four behavior controllers and 8 feedback controllers shown in Figure 8. The feedback controllers are PD controllers according the desired angle and angular velocity from the behavior controllers.

7.1   Behavior Controller

We assume that four behavior controller is needed for the robot to move from branch to branch: amplitude control behavior, approach behavior, feedback behavior and hold and release behavior.

7.1.1 Amplitude Control Behavior

This behavior controls the amplitude of oscillation based on the parametric excitation theory, changing the length of the pendulum according to the phase angle of the robot's center of gravity. We use three strategies.

(1) The elbow joint of the arm holding a branch is controlled to keep straight.

(2) The link 6 is oscillated at the same cycle as the cycle of the robot's center of gravity.

(3) The link 2, 3 and 7 are oscillated at half cycle of the cycle of the robot's center of gravity so that its length can be changed.

7.1.2 Approach Behavior

The link 6 and 7 are controlled so that the grip of free arm should come near the next branch.

7.1.3 Feedback Behavior

When the grip is near the branch, the link 6 and 7 are controlled based on the PD control in order to catch the branch,

(1) τ_{16} is adjusted to control the angle between the grip and the target branch.

(2) τ_{67} is adjusted to decrease the distance between the grip and target branch.

(3) The feedback gains are increased according to the distance between the grip and target branch.

7.2 Planner

We designed the planner which decides when and which behavior controllers should be used to make the robot locomote from branch to branch. The planner has four modes:

(1) Activate the amplitude control behavior for all links until the phase amplitude of the robot's center of gravity reaches a desired value.
(2) If the phase amplitude is in the desired range, the links 6 and 7 are controlled by the approach behavior and the other links are controlled by the amplitude control behavior.
(3) If the distance between the grip and the target branch is less than a certain value, the links 6 and 7 are controlled by the feedback behavior.
(4) When the robot's center of gravity moves away from the target branch and the branch is not in the catchable area, the link is controlled by the amplitude control behavior.

8   Simulations and Results

The hierarchical behavior controller shown in Figure 8 is applied to motion control of the seven-link brachiation robot. We designed the initial controller of body swing, to approach a target bar. The initial controller did not succeed in enabling the robot to catch the target bar in one trial but did succeed after a few swing motions as shown in Figure 9.

The evaluation functions, equations (23) and (24), are set to learn the behavior control of the body swing by reinforcement learning. The motions which are obtained in 316 learning iterations are showed in Figure 10.

$$f_m = 5 + a_1 \frac{1}{E} - a_2 \theta_{67}, \qquad \text{if robot catches the bar} \qquad (23)$$

$$f_m = \frac{1}{0.2 + \min|d^2|} + a_1 \frac{1}{E} - a_2 \theta_{67}, \qquad \text{otherwise.} \qquad (24)$$

The robot was able to catch the target bar in one trial. Each learning iteration is shown in Table 3 and Figure 11. The proposed algorithm reduced learning iterations by about 18 percent as compared with the iterations when all subcontrollers are tuned one by one.

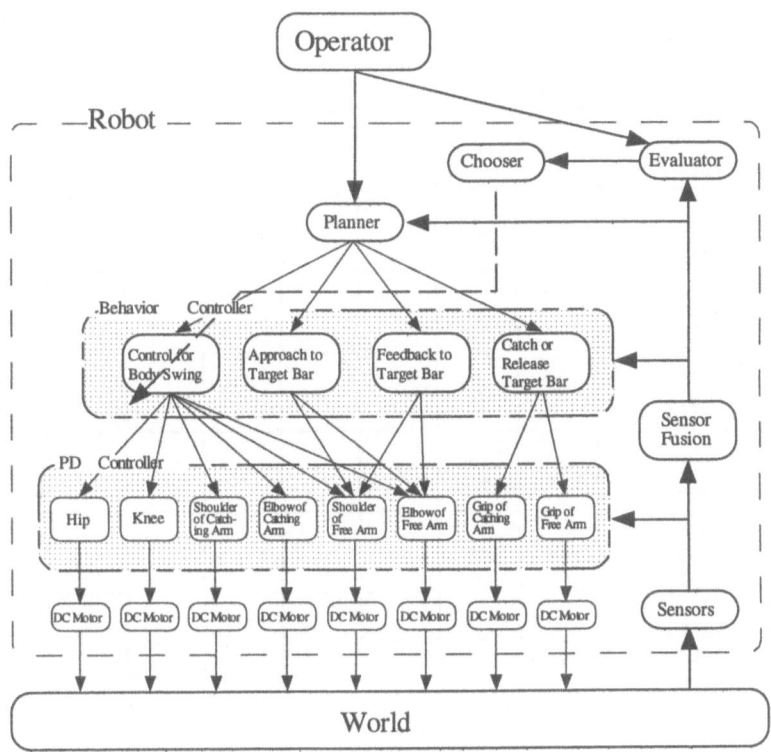

Figure 8. Control Architecture for Seven-link Brachiation Robot

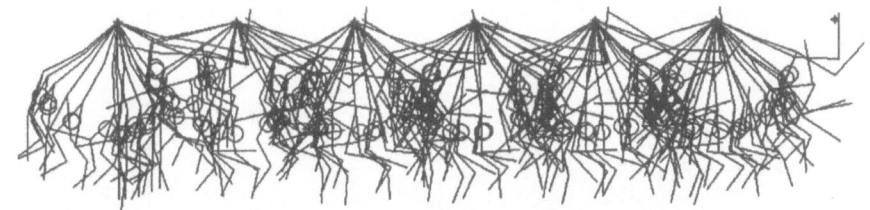

Figure 9. Stick Diagram of Seven-link Robot Locomotion
(The circle shows the center of gravity of the robot, 16ms interval)

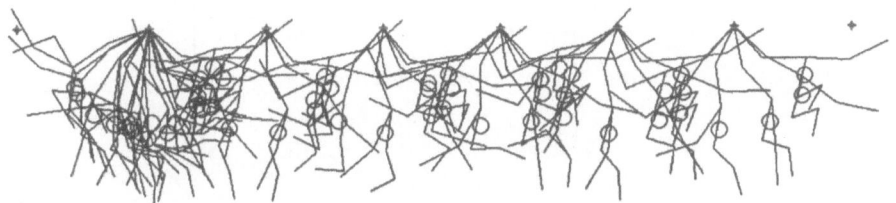

Figure 10. Stick Diagram of Seven-link Robot Locomotion
(The circle shows the center of gravity of the robot, 16ms interval)

Table 3. Change of Learning Iterations for Each Subcontroller

Learning Iterations / Subcontroller	150	158	181	245	257	265	299	316
Elbow of Catching Arm o45	25	25	33	50	54	54	62	64
Shoulder of Catching Arm o14	25	28	34	47	49	50	53	55
Hip o12	25	27	30	39	40	41	47	52
Knee o23	25	27	30	42	43	47	53	55
Shoulder of Free Arm o16	25	25	26	35	36	38	41	46
Elbow of Free Arm o67	25	26	28	32	35	35	43	44

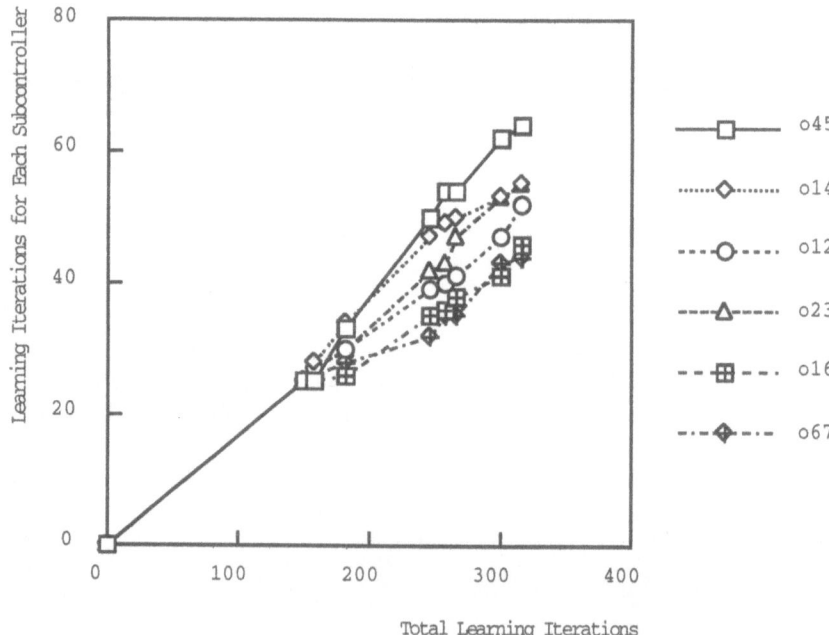

Figure 11. Learning Iterations of Each Subcontroller

9 Conclusions

We proposed the architecture of the hierarchical behavior controller and the learning algorithm to refine a behavior controller. Using these methods, we have obtained the proper behavior controller which can perform the desired behavior; the locomotion of the brachiation robot from branch and branch.

References

[1] Albus, J.S. (1991), "Outline for a Theory of Intelligence," *IEEE Trans. on Systems, Man, and Cybernetics*, Vol. 21, No. 3, pp. 473-509.

[2] Albus, J.S. (1975), "A new approach to manipulator control: The cerebella model articulation controller (CMAC)," *Trans. ASME, J. Dynamic Syst. Meas. Contr.*, Vol. 97, pp. 220-227.

[3] Fukuda, T., Saito, F., and Arai, F. (1991), "A Study on the Brachiation Type of Mobile Robot (Heuristic Creation of Driving Input and Control Using CMAC)," *Proc. IEEE/RSJ Int. Workshop on Intelligent Robots and Systems*, pp. 478-483.

[4] Shimojima, K., Fukuda, T., and Hasegawa, Y. (1995), "RBF-fuzzy system with GA based unsupervised learning methods," *Proc. 4th IEEE Int. Conf. on Fuzzy Systems/2nd Int. Fuzzy Engineering Symp. Fuzz-IEEE/IFES'95*, Vol 1, pp. 253-258.

[5] Fukuda, T., Hasegawa, Y., Shimojima, K., and Saito, F. (1995), "Reinforcement Learning Method for Generating Fuzzy Controller," *Int. Conf. on Evolutionary Computation*, Vol.1, pp.273-278.

[6] Gullapalli, V. (1990), "A stochastic reinforcement learning algorithm for learning real-valued functions," *Neural Net.*, Vol. 3, pp. 671-692.

[7] Gullapalli, V., Franklin, J.A., and Benbrahim, H. (1994), "Acquiring robot skill via reinforcement learning," *IEEE Control Syst.*, 14, pp. 13-24.

[8] Fukuda, T., Hosokai, H., and Kondo, Y. (1991), "Brachiation Type of Mobile Robot," *Proc. IEEE Int. Conf. Advanced Robotics*, pp. 915-920.

[9] Saito, F., Fukuda, T., and Arai, F. (1993), "Swing and locomotion control for two-link brachiation robot," *Proc. IEEE Int. Conf. on Robotics and Automation*, Vol.2, pp. 719-724.

[10] Spong, M.W. (1994), "Swing Up Control of the Acrobot," *Proc. IEEE Int. Conf. on Robotics and Automation*, pp.2356-2361.

[11] Paul, R.P. (1981), *Robot Manipulators*, MIT Press, pp.157.

Chapter 7

NEURAL REINFORCEMENT LEARNING FOR ROBOT NAVIGATION

S. Sehad

GRIB, University of Pierre and Marie Curie
Paris, France

Reinforcement Learning (RL) is an attractive approach for robot learning since it allows an agent to learn a given behavior from an evaluation of the wanted behavior. This measure correspond to a qualitative evaluation of the agent behavior. An agent means here a simulated system in a virtual world or a real agent interacting with a real world. The agent perceives situations of the world with its sensors and acts in the world with its motors. Figure 1 illustrates what we mean by an agent in this chapter. Experiments in learning an obstacle-avoidance behavior of the robot Khepera are presented. It is shown that neural RL is more suitable in real world applications.

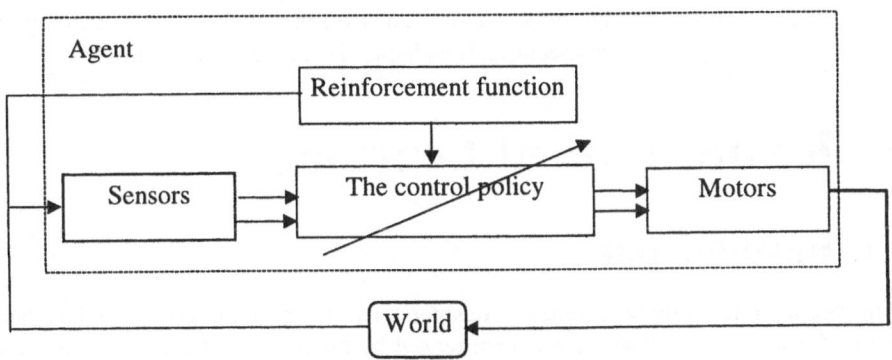

Figure 1. Functional architecture of an agent

1   Introduction

Existing RL techniques have been used by several researchers in robot learning [4][8][10][25]. But these techniques are confronted with two important problems when used in real world:

a) Memory requirement : the number of sensory inputs and actions encountered in the world occupy huge space. How can all these combinations be stored in an optimized manner ?

b) Generalization : there are so many possible combinations of sensory inputs, actions and feedback values that it is unlikely the robot will ever experience most of them. So how can it extrapolate what it has seen so far to similar situations which may arise in the future ?

Method, like those proposed by Mahadevan and Connell [13], can speed up learning but they require huge memory for storing all possible situation, action and utility value associated. Artificial neural networks are another kind of approach for RL methods. So, several implementations of the Q-learning [26] have been proposed and tested [11][25].

Section 2 presents the reinforcement learning paradigm and the Q-learning [27] algorithm. This section also reviews various refinements proposed for the Q-learning algorithm. Section 3 describes the difference between a neural implementation of reinforcement learning and a classical implementation, pointing out the advantage of the neural one. Then, our proposition is presented : a self-organizing implementation of the Q-learning algorithm. The goal of using a self-organizing map [9] is to obtain an optimal organization of the look-up table by an adaptive distribution of the entries in the neural network classification. Section 4 presents the miniature robot Khepera and the environment of our experiments. This section also describes how the obstacle-avoidance behavior and the forward moving toward a goal behavior are achieved using our proposition. Results obtained with each refinements of the Q-learning are presented and compared in this section. The chapter is concluded in section 5.

2   Reinforcement Learning

2.1   Introduction

Reinforcement learning was initially designed by the psychologists and has been studied for almost a century. It was then reused by the Machine Learning community [2][23][26]. Up until today, a clear synthetic view of the approach allowing each variation of the implementation to be positioned has not been available. By decomposing existing RL methods into functions and elements, we are able to propose a general model of the reinforcement approach. Instantiations of this general model on the widely used Q-learning [10][26] and its refinements [13] allow us to understand a neural implementation of reinforcement and to point out the advantages and disadvantages of this approach. Furthermore, we can determine where our efforts should be made to improve the performance of a neural implementation of reinforcement.

2.1.1 Definition

RL is the learning of mapping from situations to actions so as to maximize a scalar reward or reinforcement signal [25]. An agent learns a given behavior by being told how well or how badly it is performing an action which starts from a situation. It then receives a returned feedback as a single information item from the world. By successive trials and/or errors, the system determines a control policy function which is adapted through learning. For this purpose, numerous RL algorithms are available [23][26].

2.2 Methods of Reinforcement Learning

We propose to view RL as a composition of (see Figure 2) :

- — an internal state.
- — an evaluation function.
- — an update function.
- — a heuristic function.

The following elements propagate information among functions :

- — a set $X = \{x_1, \ldots, x_k\}$ of world situations. By definition, X is so large that it cannot be exhaustively explored during learning.
- — a set $A = \{a_1, \ldots, a_l\}$ of actions.
- — a reinforcement signal labeled r.

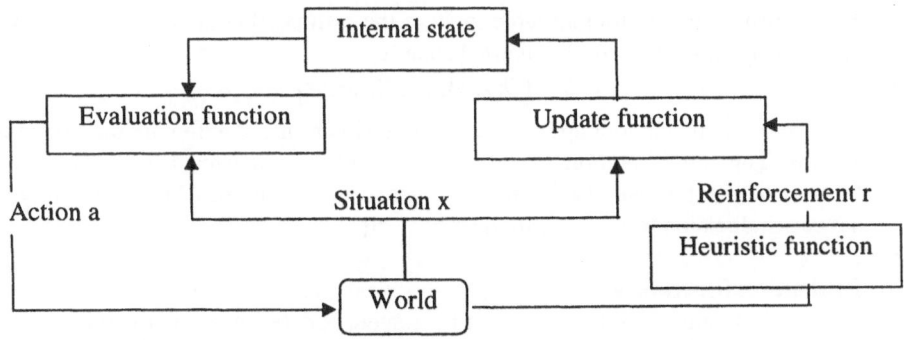

Figure 2. General decomposition of reinforcement methods

A general algorithm for reinforcement methods is then [13] :

– Initialization of the internal state S to $S_{initial}$.
  – Repeat :
    1. Let x be a world situation.
    2. Select the action a to be performed, by the evaluation function V :
       $$a = f(V(x), T)$$
       T is a random function
    3. Execute the action a in the world. Let r be the immediate reward (if it is available) associated with the execution of the action a in the world.
    4. Update the internal state S by the update function U :
       $$S_{new} = U(S_{old}, x, a, r)$$
       Knowledge about returned rewards is stored in S.

The heuristic is a function given by a human expert. It is a formula which leads the agent to adopt the function f representing the behavior . It is the only function dependent on the application. The reinforcement r is a qualitative signal (i.e. 1 : good ; -1 : bad ; 0 : unknown) returned by the heuristic function H.

The Q-learning is the most known and used algorithm in reinforcement learning. We propose a description in the formalism given above for each of the following reinforcement algorithms : Q-learning and its refinements (Q-learning with Hamming distance, Q-learning with statistical clustering and Dyna-Q).

2.2.1 Q-Learning Algorithm

The Q-learning algorithm builds a Q function that maps situation-action pairs (x, a) into expected returns r. $Q(x, a)$ is the system's estimate of the return it expects to receive given the fact that it executes action a in situation x.

a) **The internal state S** :
   The algorithm uses a lookup table to store the estimated cumulative evaluation Q. All these values represent the internal state S.
   $$S = \{Q(x, a), x \in X, a \in A\}$$
   Row indices in the lookup table represent situations, column indices represent actions. The representation of the internal state does not need to be only in the form of tables, but can also be more compact representations like artificial neural networks, decision trees or symbolic rules [25].

b) **Evaluation function V** :
   Using the Q values, the behavior π of the agent is determined by the rule :
   $$\pi(x) = a \text{ such that } Q(x_t, a) = \max_{b \in A} Q(x_t, b).$$

Where a represents the best utility action to perform giving the current situation x_t. The really performed action is also dependent on the exploration function used. The best exploration strategy must take into account the utility action values. For ameliorating the agent behavior this function must take in care actions with high utility value. The most used function is based on the Boltzmann distribution[27].

$$P(a/x_t) = \frac{\exp \dfrac{Q(x_t,a)}{T}}{\displaystyle\sum_{b \in A} \exp \dfrac{Q(x_t,b)}{T}}$$

c) **Update function U :**
   The internal state Q is updated by the following function :

$$U: Q_{new} = U\left(Q_{old}, x_t, a, x_{t+1}, r\right) \text{ or}$$

$$Q(x_t, a)_{new} = Q(x_t, a)_{old} + \beta.\left(r + \gamma.\max_{b \in A} Q(x_{t+1}, b) - Q(x_t, a)_{old}\right)$$

where β and γ are constant coefficients, between $0 < \beta, \gamma < 1$.

The reinforcement at the present time should be equal to the expected returned rewards. The error between the expected value $r + \gamma.\max_{b \in A} Q(x_{t+1}, b)$ and the current value $Q(x_t, a)$ must then be minimized. This updating rule has the effect of propagating a reward associated with a given situation-action. It is, in fact, a way to backpropagate delayed rewards (see Figure 3)

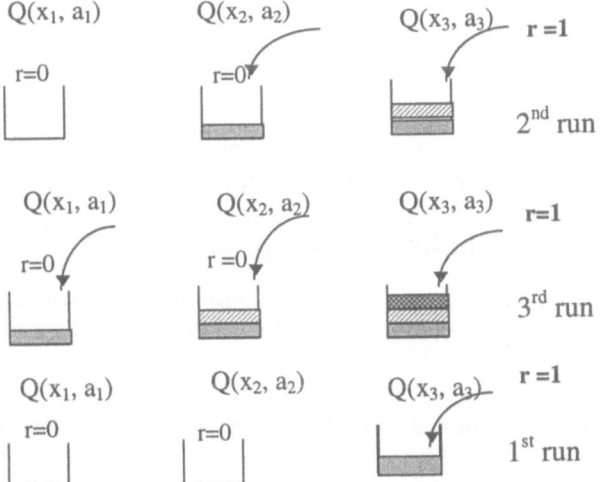

Figure 3. Example of backward propagation of a delayed reinforcement on 3 runs of the same sequence of situation-action.

2.2.2 Q-Learning with Weighted Hamming Distance

The main idea of this refinement is to learn faster. With this in mind, Mahadevan and Connell [13] propose computing a Hamming distance between the real situation x and similar situations in order to apply the update function on all of them and for the same reward.

a) Update function U :
The update function U (see Figure 4) is the same as the one used in the Q-learning. However, all similar situations are updated at the same time using the same reward. The Hamming distance between any two situations is simply the number of bits that are different between them. Bits can be of different weights. Two situations are distinct if the Hamming distance between them is greater than a fixed threshold ρ.

Example : $x_2=(0001)$, $x_3 = (0011)$, ρ=2 ;   Hamming distance(x_2, x_3)=1<2.

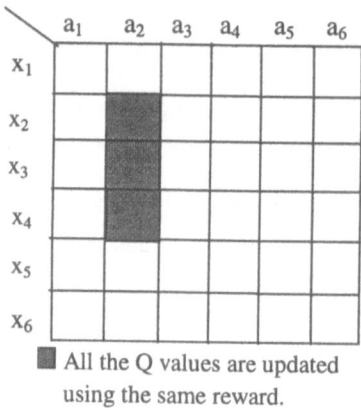

■ All the Q values are updated
using the same reward.

Figure 4 : Updated Q values for similar situations (in the sense of Hamming distance) after executing a_2 in x_3

2.2.3 Q-Learning with Statistical Clustering

The goal here is the same as in the previous approach, Mahadevan and Connell [13] use statistical clustering to propagate returned rewards across situations.

a) Update function U :
In order to propagate a returned reward for a situation x, the same function U that was used in Q-learning is used here.
Each action is associated with a set of clusters giving information concerning the usefulness of performing the action in a particular class of situations. Clusters are a set of « similar » situation instances which use a given similarity metric. Each situation x' is updated, if it belongs to a cluster in which x appears (see Figure 5)

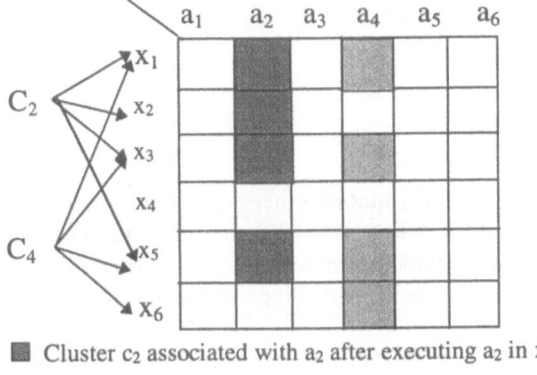

■ Cluster c_2 associated with a_2 after executing a_2 in x_3

▣ Cluster c_4 associated with a_4 after executing a_4 in x_3

Figure 5 : Updating of the Q values associated with 2 clusters.

2.2.4 Dyna-Q

It may be difficult to return the same sequence of situations-actions to backpropagate delayed rewards. As a result , Sutton [24] added a model of the world in which pairs of situation-action are randomly carried out again (see Figure 3).

a) **Heuristic function H** : The H function is modified in order to deal with the modeled world. In this case, only previously seen pairs of situation-action (in the real world) will lead to a non-zero reinforcement reward. The returned reward r' is the same as in the real world.

b) **Evaluation function V** : When the experience is performed in the real world, V is the maximum function. Otherwise (i.e., for an experience in the modeled world) V is a random function.

c) **Update function U** : The updating rule uses rewards r or r' indifferently.

2.3 Limitations of Classical Reinforcement Learning Algorithm

As we have seen, RL methods are faced with two important problems. Generalization is limited to syntactic criteria. Methods like those proposed by Mahadevan and Connell [13] and Sutton [25], can speed up learning but they are nevertheless subject to a memory requirement for storing all possible situation-action utility values. Neural networks are another kind of approach for RL methods.

3 Neural Reinforcement Learning

3.1 Introduction

Numerous authors ([1][3][7][15] among others) propose a neural implementation of reinforcement learning, but it is not clear what the differences are, nor where they lie compared with a classic implementation. In our formalism, a neural network implementation of RL implicates the following modifications:

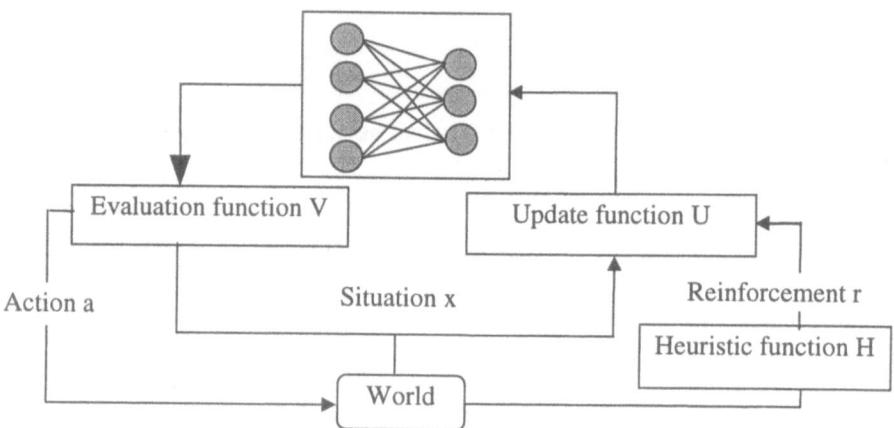

Figure 6 The internal state represented by a neural network

a) **The internal state S** :
Instead of representing the internal state by a look-up table, it is represented by a neural network. The internal state is constituted by the weight set of the network (W). The memory size required by the system to store the knowledge is then defined, a priori, by the number of connections in the network. It is independent of the number of explored situation-action.

b) **Evaluation function V** :
The evaluation is the result of the processing by the network of the input situation x plus a random component α_t. This component decreases during the learning process. At the same time, the network generalization becomes better : the system learns. The selection of an action is not based on deterministic quantities, like Q values. The neural network proposes « the best » action to perform in the given situation, but it gives no information about the expected reinforcement or other actions having the same usefulness. It should be remembered that in classic RL methods, each utility value associated with a situation-action pair is stored and remains available for computation. The architecture of the neural network can be multilayer if mapping from situations to actions is complex.

c) **Update function U** :
The update function U works on the internal state. When applied to a neural network, it is a weight modification algorithm. Qualitative reinforcement gives information about how well the system behaves In the case of positive reinforcement signal, it is particularly easy to determine the output error. This error is equal to the added random values. Therefore, the gradient error descent algorithms are good choices for updating functions. As regards to negative reinforcement signal, an easy definition of the output error is restricted to simple cases where only two actions are possible. In this case, the desired output is the other action. If actions are binary coded, the desired output is then the inverse of the network proposal. As we can see, the definition of an error in the case of negative reinforcement is difficult. It is important to note that this update rule is not the same as for Q-learning.

3.2 Discussion

Among the advantages of neural reinforcement learning are a limitation in the memory requirement and more intelligent exploration of the situation-action space. The generalization achieved by the neural network cannot be characterized as easily as for the Hamming distance or clustering techniques. Nevertheless, all connectionist applications used in industry serve to highlight the interest one can find in using artificial neural networks. The main limitation which arises when using neural reinforcement method is that the reward value associated with a situation-action is not available.

3.3 Self-Organizing Map Used for a Neural Implementation of Q-learning

Up until today, neural implementation of the Q-learning algorithm represent the association between the sensory input space and the action space using a multilayer perceptron (MLP) [19]. We propose to use the self-organizing map [9] (SOM) to represent the mapping between these two spaces. The self-organizing map is distinguished by the development of a significant spatial organization of the layer. This neural network develops internal representations of the input space by mapping distinct input vectors into distinct areas of the network layer. Weights specify clusters that sample the input space such that the point density function of the cluster tends to approximate the probability density function of the input vectors. In addition, the weights are organized such that topologically close neurons are sensitive to inputs that are physically similar.

We use the self-organizing map as an adaptive look-up table [17][18] of situations, actions and utility value where each unit i in the SOM codes the sensory input x, the action a as well as the utility value Q respectively by the weight vectors w_x^i, w_a^i, w_Q^i.

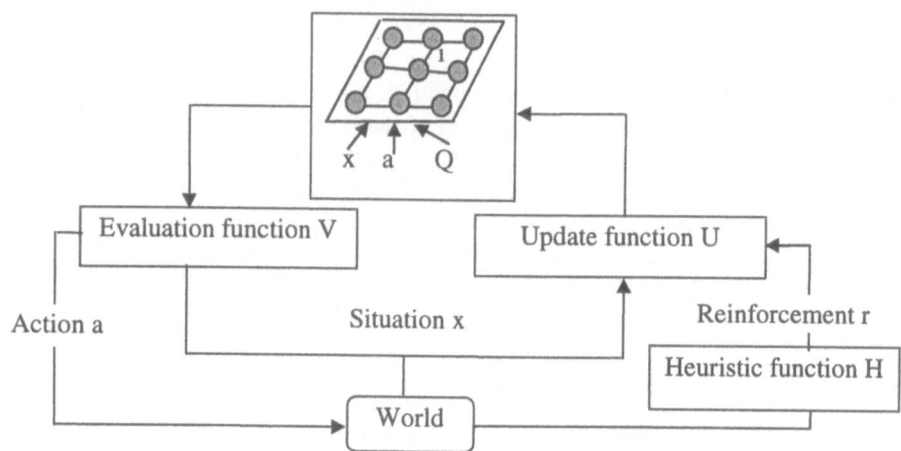

Figure 7 : The internal state is represented by a self-organizing map.

a) **The internal state** :

The internal state is composed of set of weights of the network (W). Inputs to the network are composed of three components. Weights $\left(w_x, w_a, w_Q\right)$ of each neuron of the layer store the three input components. The memory size required by the system to store the knowledge is defined, a priori, by the number of connections in the network. It is independent of the number of explored situation-action pairs.

b) **Evaluation function** :

The SOM is then used as an associative memory. It is therefore possible to retrieve information using an incomplete probe. Let use ask the following question (see Figure 8) [21]: for a given sensory input x of the robot in the workspace, what is the best rewarding action a to be taken in order to get the best reward value?

The probe is therefore the current situation x_t, and the maximum utility value $Q = 1$ (normalization of the weights allows us to put this maximum value to +1).

Response : (........, w_a^i , w_Q^i)

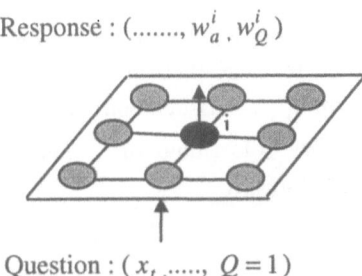

Question : (x_t ,....., $Q = 1$)

Figure 8 : During the evaluation phase, the input vector is only composed by the sensory input x_t and the best utility value that can be associated $Q = 1$. Then, we determine the winner neuron i which the weights vector is the closest to the input vector. This weights vector provides us with the action value a_t and the utility value Q_t really associated with the situation-action pair (x_t, a_t).

The activated neuron corresponds to a vector composed of situation, action and utility value. The action retrieved from the self-organizing map is the one which offers the best reward in the present situation. A random component is added to the proposed action so as to explore the search space.

• <u>Neuron selection</u> :
The selected neuron i is the one with the minimum distance. We compute the distance d_j between the input and each neuron j using

$$dj = \left| x_t - w_x^i \right| + \left| 1 - w_q^i \right|$$

where
x_t is the current situation, submitted to the map,
1 is the best possible utility that can be associated with x_t.

• <u>The action which is performed by the agent</u> :
In the situation x_t the action a_t to be performed by the agent is given by :

$$a_t = \alpha_t . + w_a^i$$

where
w_a^i is the weight coding the action of the selected neuron i .
$0 \le \alpha_t \le 1$ is a variable parameter which decreases with the learning steps.

• <u>The utility value Q really associated with the pair</u> (x_t, a_t) :
The utility value Q_t is retrieved :

$$Q_t = w_Q^i$$

where

w_Q^i is the weight coding the utility of the selected neuron i .

c) **Update function** :

Once a neuron has been selected, its weights and those of the other neurons in its neighborhood are modified to make these neurons more responsive to the input (x_t, a_t, Q_t). Before learning this association (x_t, a_t, Q_t), we must update the utility value Q_t with the Q-learning rule. This update is necessary for backpropagating the delayed reward.

Response : $(..,..., w_Q^j)$

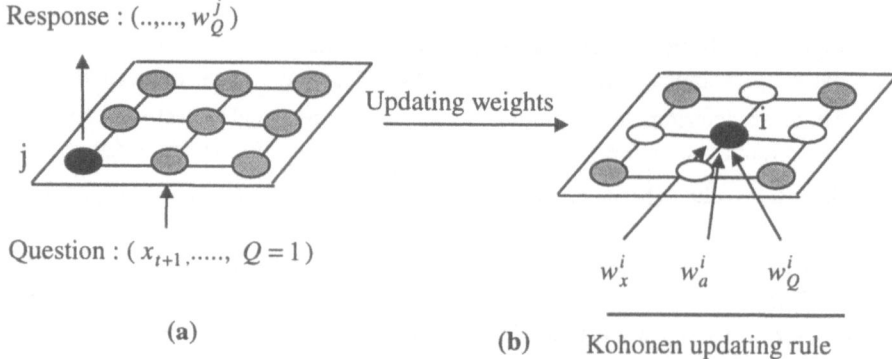

Updating weights

Question : $(x_{t+1}, ..., Q = 1)$

$w_x^i \quad w_a^i \quad w_Q^i$

(a)

(b)   Kohonen updating rule

Figure 9 : (a) x_{t+1} is the situation after the agent performs the action a_t in the situation x_t. The input vector $(x_{t+1}, ..., Q = 1)$ is presented to the map and permitted to determine the best utility value associated with the situation x_{t+1}. (b) Then, the weights (w_x^i, w_a^i, w_Q^i) are updated using the Kohonen updating rule.

The following steps describe the updating phase :

1. Update the utility value Q_t associated with the pair (x_t, a_t) according to the Q-learning rule :

   But first, determine the best utility value Q_{max} associated with the next sensory input x_{t+1}. Select the closest neuron j by presenting to the map the input vector $(x_{t+1}, ..., Q = 1)$ (see Figure 9 (a)) :

   $$Q_{max} = w_Q^j$$

   then

   $$Q_t = Q_t + \beta(r + \gamma \cdot Q_{max} - Q_t)$$

   $0 \le \beta \le 1$ and $0 \le \gamma \le 1$ are constant parameters.

2. Update the weight vectors of the neuron i (and its neighbors k) :

    The neuron i is the one determined at the evaluation phase. The weights adaptation is performed for the vectors w_x^i, w_a^i, w_Q^i using the Kohonen rule (see Figure 9 (b)):

    $$w_z^i = w_z^i + \mu\left(z - w_z^i\right), \text{ for the neuron } i$$

    $$w_z^k = w_z^k + \nu\left(z - w_z^k\right), \text{ for the neighbors neurons k,}$$

    where z correspond to x_t, a_t or Q_t and $0 \le \mu \le 1$, and $0 \le \nu \le 1$ are gain parameters which decrease with the learning steps.

3.4 Discussion

The self-organizing implementation of the Q-learning responds to the two problems faced by reinforcement learning: the memory requirement and generalization.

*   The memory requirement : Each neuron represents a cluster of situation, action and a utility value. By comparison with a look-up table where every association is stored, the map allows to store just representative associations of them, so the memory requirement is resolved by this way.

*   Generalization : Each neuron on the map codes a cluster of vectors (situation, action, utility). Updating the utility value associated with a (situation, action) is performed for all situations belonging to the same cluster.

4   Experiments with Khepera

4.1 The Mini Robot Khepera

Khepera [16] (see Figure 10) is a miniature robot of circular shape measuring 55mm in diameter, 30mm in height and 86g of weight. It was developed to test mobile robot control algorithms in the real world. Its configuration consists of two wheels and 8 infrared sensors placed around its body (6 on the front, 2 on the back). These sensors are sufficient for simple obstacles and light detection. Sensor values range from 0 (for a distance > 5 cm) to 1023 (approximately 2cm). Wheels have integer values between]-10, 10]. Khepera can be used on a desk connected to a workstation through a wired serial link. This configuration allows an optimal experimental configuration with everything at hand : the robot, the environment and the remote host computer.

Figure 10. Khepera, the miniature mobile robot [16]. Its configuration consists of two wheels and 8 infrared sensors placed around its body (6 on the front, 2 on the back)

4.2 The Arena

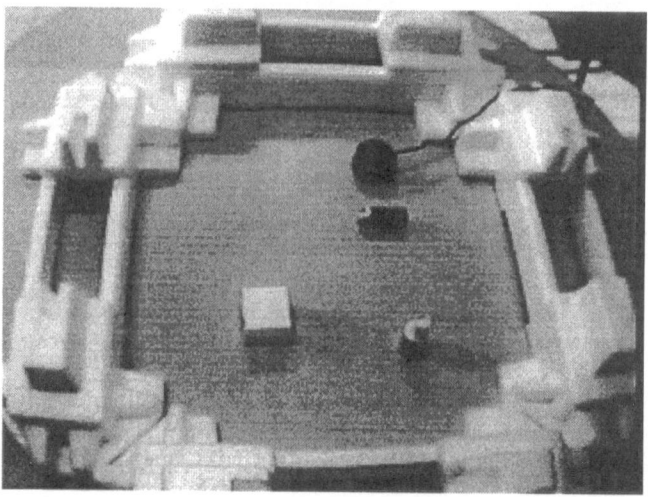

Figure 11. Khepera inside its environment. The robot diameter is 5.5 cm. The arena dimensions are 40 cm x 40 cm. Obstacles are a little box, eraser, and a Tipp-Ex box.

The robot is put in a 40 cm x 40 cm maze (see Figure 11) with lighted shaped walls. Obstacles with different shape, form and color (Little box, Eraser, Tipp-Ex box) are introduced at different places in the maze.

4.3 Comparison: Q-learning Versus Neural Q-learning

We have compared performance of the Q-learning algorithm and its refinements with the neural implementation we propose for learning a simple task : obstacle-avoidance. Our goal is not to solve a problem for which solution already exist, but to work with a complex problem to allow a good understanding of the need and implications of neural implementations.

4.3.1 The Robot Task: Obstacle-Avoidance Behavior

a) **The heuristic function**:
The heuristic function has been subject to many modifications. For the learning of an obstacle-avoidance behavior, we compare the past and the present sensor values. If there is more excitation than the last measurement then r = +1 (it is good), otherwise if the present sensor values presents a level which is too high then r = -1 (it is bad)

Obstacles-avoidance heuristic $\left(x_i(t-1), x_i(t)\right)$

Let $x_i(t)$ be the sensor value of the sensor i at time t and r the reinforcement signal :

$$\text{if } \left(\left(\sum_{i=1}^{8} x_i(t-1) - \sum_{i=1}^{8} x_i(t)\right) \geq -0.0073\right) \text{ than } r = +1$$

$$\text{else} \quad \text{if } \left(\left(\sum_{i=1}^{6} x_i(t)\right) > 0.366\right) \text{ or } \left(\left(\sum_{i=7}^{8} x_i(t)\right) > 0.122\right) \text{ than } r = -1$$

$$\text{else } r = 0$$

Threshold values like (0.00732, 0.366, 0.122) have been determined after extensive experimentation. 0.366 are set for the front sensors and 0.122 for the backward sensors.

b) **The internal state :**
For the same task to be performed by the robot, we compare the (situation x action) space for every implementation :

Table 1. Comparison of the situation space, the action space and the combined space used with the Q-learning and its refinements.

	Q-learning	+ Hamming	Dyna-Q	+ Clustering	+Kohonen
Situation space (X)	$2^8 = 256$			2^{16}=64K	$1024^8 \approx 10^{24}$
Action space (A)	$5\text{x}5 = 25$				20x20=400
(X x A) space	6400			1.6M	4.10^{26}

For the classical implementation of Q-learning, we are obliged to reduce the (situation x action) space by a binary coding (0 or 1) of the 8 sensors. For the Q-learning, Q-learning with Hamming distance and Dyna-Q, each sensor is coded by one bit. For the Q-learning with clustering, each sensor is coded by 2 bits.

With the self-organizing map implementation, the size of the situation-action space is 1024^8 x (20x 20), about 4.10^{26} . Dimension of the self-organizing map is 2 (4 neighboring neurons) and 16 neurons are on the map. The input to the map is a vector of 11 bits (8 sensors value , 2 actions, 1 utility). The size of the weights matrix is (16 x 11). The weights matrix is initialized randomly at the beginning.

4.3.2 Experiments and Results

We place the robot in a maze, where obstacles can be introduced or removed. The exploration phase allows Khepera to encounter different spatial situations and to associate an action and a utility for each situation. Several experiments helped us determine the best parameters values for the Q-learning and the Self-organizing map. Constant parameters of the Q-learning algorithm β and γ are set respectively to 0.50 and 0.9. Constant parameters of the Kohonen algorithm μ and ν are set to 0.7 and 0.3 (for the neighbor neurons).

During the learning phase, randomness is gradually reduced towards zero. Figure 12 shows performance comparison during the learning phase with the Q-learning (Q) algorithm and its variants (Hamming (H), Statistical Clustering (C), Dyna-Q (D) and Kohonen self-organizing map (K)) for the obstacle-avoidance behavior. The robot exhibits a correct behavior after about 600 steps with the self organizing map (see Figure 12). With Q-learning using weighted Hamming distance, the robot exhibits a correct behavior after about 4000 steps.

The generalization ability of the behavior synthesized is illustrated by removing or adding obstacles in Khepera's spatial environment. Our criterion for having learnt is no more collision occur estimated subjectively by the experimenter, together with a discounted   cumulative   reward.   This   discounted   cumulative   reward
$$\left(\frac{\sum r_{all} - \sum r_{-}}{\sum r_{all}} \right) \leq 1$$ corresponds to a measure of the cumulated performance of the robot. It is the number of good actions (r = +1) performed per the total number of reward obtained.

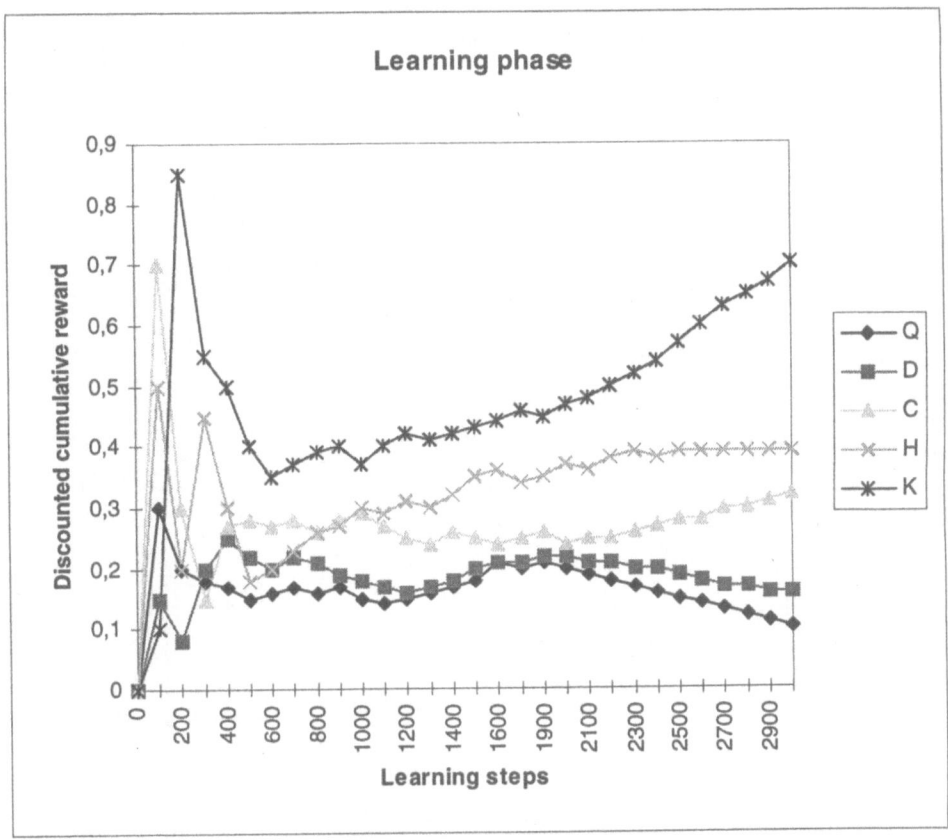

Figure 12 : Performance obtained during the learning phase with the Q-learning (Q) algorithm and its variants (Hamming (H), Statistical Clustering (C), Dyna-Q (D) and Kohonen self-Organizing map (K)) for the obstacle-avoidance behavior.

Table 2 compares the number of learning steps and Learning time for each implementation of the Q-learning. The learning time is the time in minutes needed to synthesize an obstacle-avoidance behavior.

Table 2. Comparison of the number of learning steps and Learning time for each implementation of the Q-learning. The learning time is the time in minutes needed to synthesize an obstacle-avoidance behavior

	Q-learning	+ Hamming	Dyna-Q	+ Clustering	+Kohonen
Learning Steps	7500	3500	6000	4000	600
Learning Time	55 mn	25 mn	45 mn	30 mn	5 mn

Figure 13 shows the performance obtained during the test phase of the Q-learning (Q) algorithm and its variants (Hamming (H), Statistical clustering (C), Dyna-Q (D) and

Kohonen self-organizing map (K)) for the obstacle avoidance behavior after 3000 learning steps.

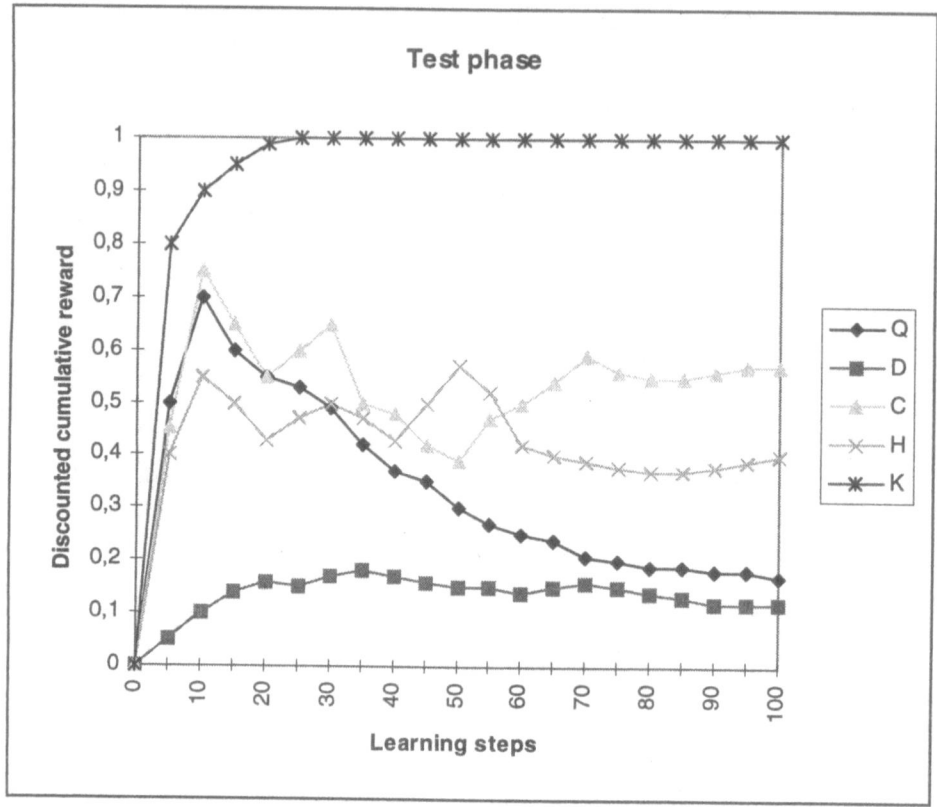

Figure 13. Performances obtained during the test phase of the Q-learning (Q) algorithm and its variants (Hamming (H), Statistical clustering (C), Dyna-Q (D), Kohonen self-organizing map (K)) for the obstacle avoidance behavior after 3000 learning steps.

4.4 Kohonen Weights Interpretation

The interpretation of the learning behavior from the network weights is possible. In Figure 14 and Figure 15, we present the weights after 5000 learning steps. These weights depend only on the explored situation (respectively action) and represent for each neuron the prototype situation (respectively action) associated. Figure 14 presents the 16 prototypes of situation clusters by each of the 16 neurons (4x4) of the map. Figure 15 presents the corresponding 16 prototypes action clusters learned by the same 16 neurons.

For example, the diagram for the neuron number 4 of Figure 14 shows that this neuron is associated with a situation presenting an obstacle on the front: principally coded by sensor 3, but also to a lesser level by sensors 1, 2, 4 and 5. This obstacle must be at a distance of approximately 3 cm.

The diagram for the neuron number 4 of Figure 15 shows that this neuron is associated with an action associated to the previous situation. Values represent speed of the right and left motors. Note that the two speed values are negative values. Speed value of the right motor is greater than the speed value of the left motor. The robot performs a backward motion turning to the left side. This action allows it to avoid the obstacle not far in the front.

Note also that the number of prototypes learned depend only on the number of neurons used on the map. It does not depend on the position of the obstacles in the arena. A disadvantage of the proposed method is the selection of the size of the Kohonen network. It would be worthwhile to use a neural network where the number of neurons can be adapted [5].

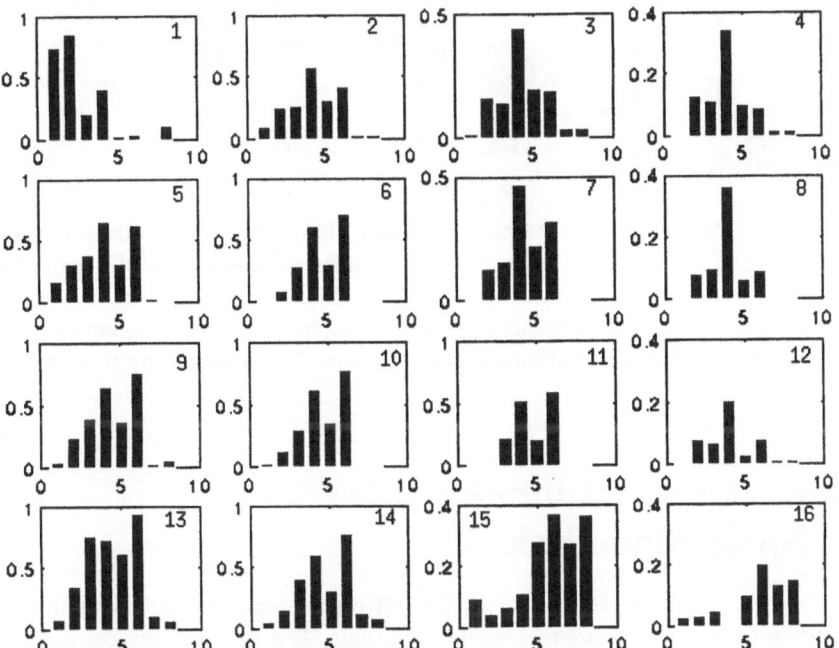

Figure 14. Sensorial situations learned by a 4x4 Kohonen map. Visualization of the eight weights associated with the sensory input for each neuron of the map after 5000 learning steps in a task of obstacle avoidance behavior. The higher the value of the sensor, the more sensitive the corresponding neuron to an obstacle.

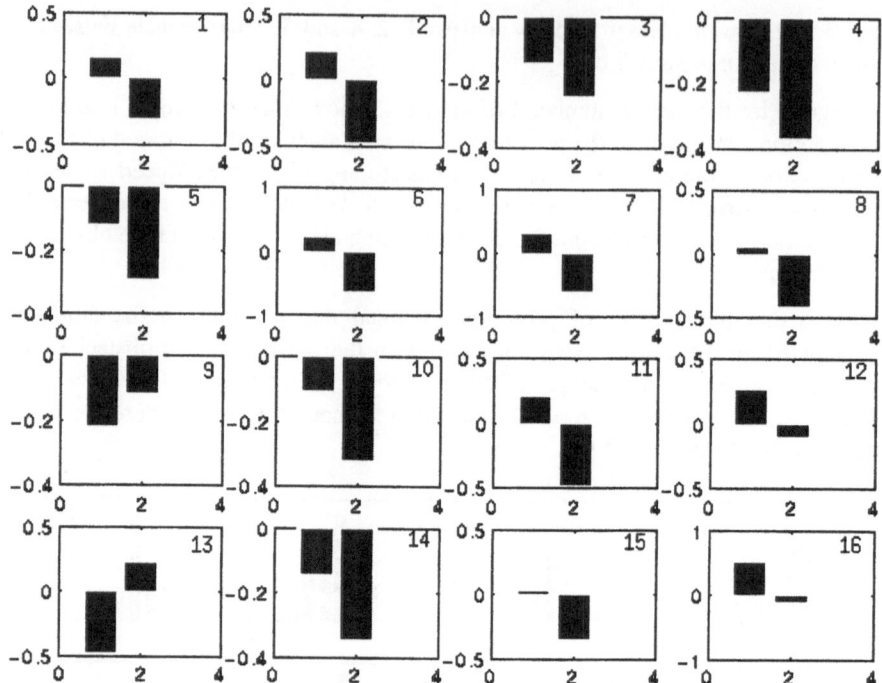

Figure 15. Action clusters learned by a 4x4 Kohonen map. Visualization of the two weights associated with the action coding for each neuron of the map after 5000 learning steps in a task of obstacle avoidance behavior. Each action is associated with a class of sensory input The weights values correspond to speed values associated with right motor and left motor. These values are positive and negative.

4.5 Learning a Complex Behavior: The Navigation Task

In this section, we illustrate how we use the self-organizing map implementation of the Q-learning for learning a more complex behavior. The task to perform is to move forward when possible to reach a goal and avoid obstacles. In our experiments, this navigation task is decomposed into two separate behaviors [12][14][20]:

- an obstacle-avoidance behavior
- and a forward moving toward a goal.

Figure 16 illustrates this decomposition. Since avoiding obstacles precede moving toward a goal, this behavior is given the higher priority and is always active. The obstacle-avoidance behavior used here is acquired previously.

Obstacle-Avoidance

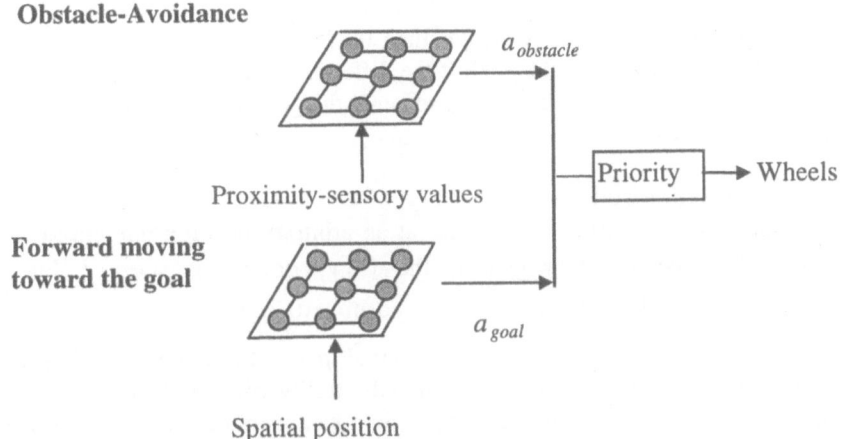

Proximity-sensory values

**Forward moving
toward the goal**

Spatial position

Figure 16. Decomposition of the navigation behavior : Two self-organizing maps associated with the Q-learning for learning an obstacle-avoidance behavior and a forward moving toward a goal behavior. The two behaviors are then composed by priority .

4.5.1 Forward Moving Toward the Goal

To learn this behavior , the following question should be asked [22]:

« For a given position and orientation of the robot in the workspace, what is the best rewarding action a to be taken in order to reach the goal (by the shortest path) ?».

The probe to the map is therefore the current position, the current robot orientation, the goal position and the maximum utility value Q ($= 1$). The activated neuron corresponds to a vector composed of a position, orientation, goal position, action, utility value. The action retrieved from the self-organizing map is the one which offers the best reward in the present situation. A random component is added to the proposed action so as to explore the search space.

a) **The map (the internal state)** :
   In our experiments, we used a two-dimensional map with 256 neurons (16x16). The Kohonen map receives an input vector of 9 bits (current position (x_t, y_t), direction (d_x, d_y), the goal position (x_g, y_g), the action (a_x, a_y) and the utility value Q. Positions are simply pairs of Cartesian coordinates. The position and direction of the robot in the workspace are given by a computed odometry. The size of the weights is (256 x 9). They are initialized randomly at the beginning.

b) **The heuristic function** :

We compute a Manhattan distance between the goal position and the next position (x_{t+1}, y_{t+1}) and the goal position and the last position (x_t, y_t). If the difference of the two computed distances is greater than 0.02 then r= -1. This means that the robot is moving away from the goal. If the difference is less than 0.02 then r = +1. This indicates that the robot is moving towards the goal. 0.02 has been determined through experiments.

c) **Experiments and results** :

The robot is put in a 40 x 40 cm maze at an arbitrary position referenced as the position (0,0). An arbitrary goal in the (x, y) domain is selected. For this experiment, the goal is set to the cartesian position (0.7, 0.7).

The learning phase allows Khepera to encounter pairs of different spatial positions and directions and to associate an action and a utility with each pair, relative to a goal position. Obstacle avoidance is made by a Kohonen map associated to the Q-learning. We use the previous learned map for this behavior. Through several experiments, the gain parameters associated to the map are setting at 0.34 for the winner neuron and 0.30 for its neighbors. Constant parameters β and γ associated with the Q-learning are setting to 0.6 and 0.4.

After about 5000 learning steps we test the performance of the robot. Khepera is placed again at the position (0, 0). When moving it is attracted by the goal and don't collide obstacles. The robot does not stop exactly at the goal position but continues to move around it. Performance of Khepera shown in Figure 17 and Figure 18 display discounted cumulative reward over time. The graphs measure the performance of the algorithm in minimizing errors.

Figure 17 shows performance obtained by Khepera during the learning phase. Discounted cumulative reward is calculated through 5 experiences. During this phase, the performance is variable due to the exploration function. Performance obtained in the beginning depends on the initial position of Khepera (placed front of the goal).

Figure 18 shows three different related experiences. It is evident that Khepera reach the goal in about 300 steps (Experience 1 : 270 steps, Experience 2 : 360 steps, Experience 3 : 300 steps). During this phase, actions performed by Khepera are those proposed by the map.

The method builds a path in which every action taken by the robot leads to the best future rewards. Positive rewards are given by means of minimizing a Manhattan distance between the current position and the goal position. The generated path takes into account the location of the obstacles encountered during the learning phase. Note that none of the neurons encode an obstacle location. An interesting question is : is the path finally found, after ≈ 5000 steps, the one having the lowest cost ?

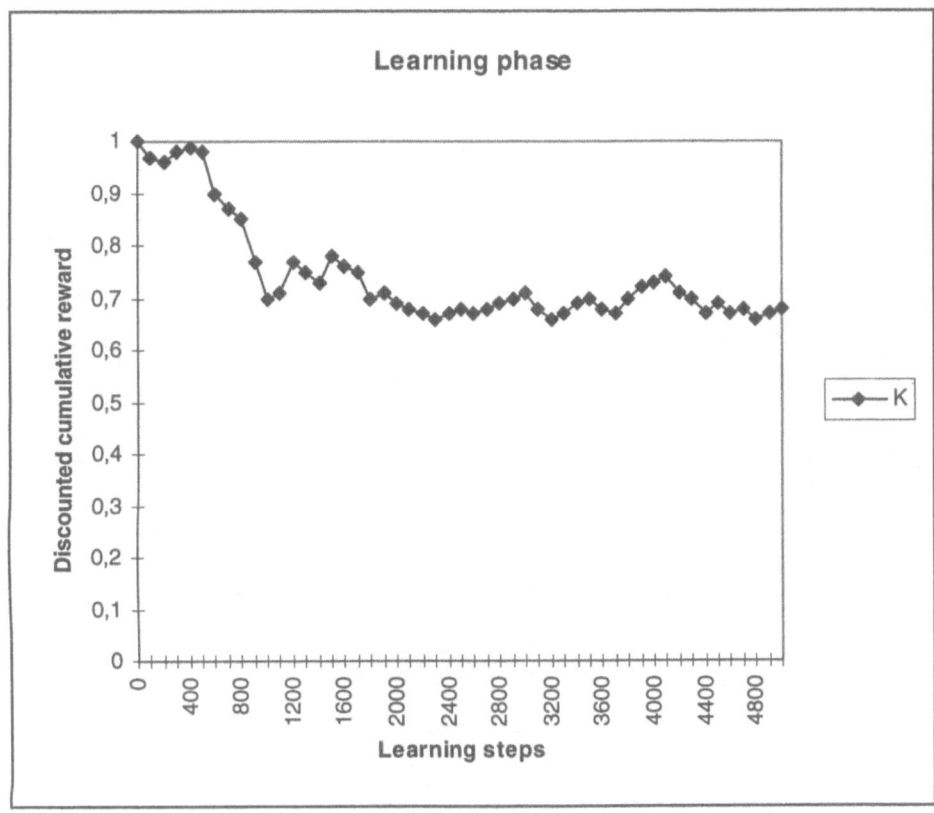

Figure 17. Performances obtained by Khepera during the learning phase. Discounted cumulative reward on 5 experiences. During this phase, the performance is variable due to the exploration function. Performances obtained in the beginning depend on the initial position of Khepera.

5   Conclusion

In this chapter, we describe the implementation of the self-organizing map of the Q-learning algorithm. We show the use of the Kohonen implementation of Q-learning to solve the two problems confronted by RL. Generalization is a well known property of the self-organizing map. The memory requirement is limited to the number of connections on the map. Experiments in learning an obstacle-avoidance behavior with the robot Khepera illustrate the efficiency of this neural implementation in a situation-action space of considerable size compared with other refinements like the Q-learning with Hamming distance, with clustering and Dyna-Q. We have also shown how we use this neural implementation to learn a more complex task: a navigation behavior. Our aim was not to resolve these well known tasks but just to illustrate our implementation. As shown by the results, this way of using neural RL appears applicable in real world applications.

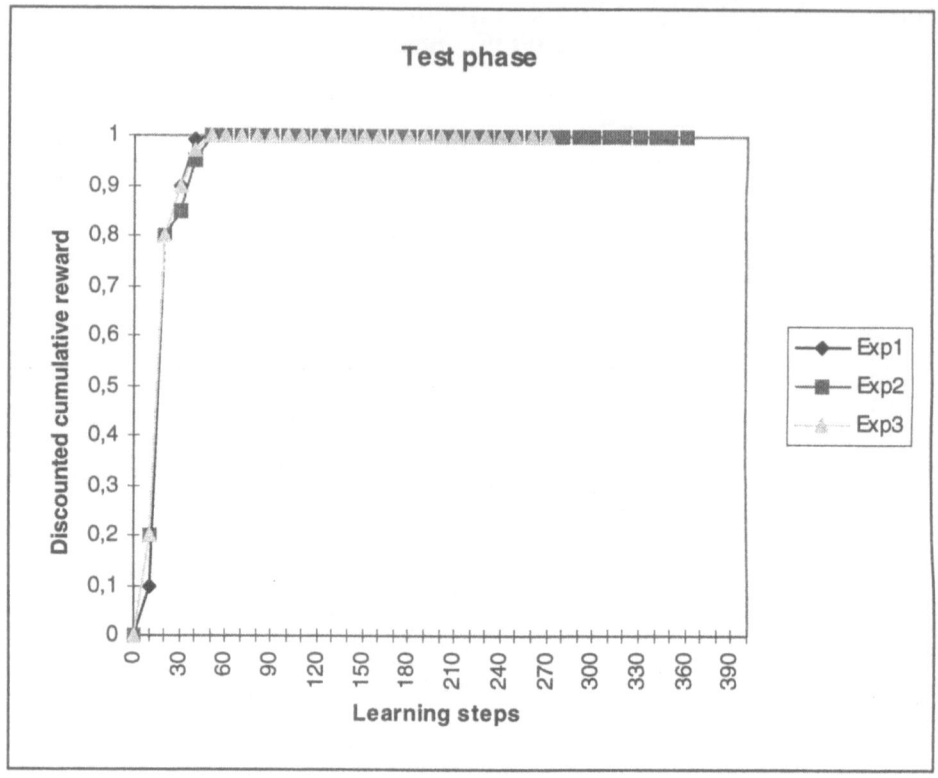

Figure 18. Performance obtained by Khepera during the test phase. Discounted cumulative reward on 3 experiences. During this phase, actions are those proposed by the self-organizing map.

Acknowledgments

We thank Francesco MONDADA (K-TEAM, EPFL-LAMI, Switzerland) for the use of one of the first Khepera robots.

References and Further Reading

[1]   Anderson, C.W. (1988), "Learning to Control an Inverted Pendulum Using Neural Networks," *American Control Conference*, Atlanta, Georgia, USA.

[2]   Barto, A.G., Sutton, R.S., and Anderson, C.W. (1983), "Neuronlike Elements That Can Solve Difficult Learning Control Problems", *IEEE Transactions on Systems, Man and Cybernetics*, (13), pp. 834-846.

[3]   Barto, A.G. (1992), "Reinforcement Learning and Adaptive Critic Methods," *Handbook of Intelligent Control: Neural, Fuzzy, and Adaptive Approaches*, Edited by D.A. White and D.A. Sofge, pp. 469-491.

[4]   Dorigo, M. and Collombetti, M. (1994), "Robot Shaping: Developing Autonomous Agents through Learning," *Artificial Intelligence*, Vol. 71, (22), pp.321-370.

[5]   Fritzke, B. (1993), "Growing Cell Structures - a Self-organizing Network for Unsupervised and Supervised Learning," *Technical Report*, TR-93-026, ICSI-International Computer Science Institute, Berkley, California, USA.

[6]   Gullapalli, V. (1994), *Reinforcement Learning and Its Application to Control*, Ph.D. Thesis, University of Massachusetts.

[7]   Hertz, J., Krogh, A., and Palmer, R.G. (1991), *Introduction to Theory of Neural Computation*, The Advanced Book Program.

[8]   Kaelbling, L. (1993), *Learning in Embedded Systems*, MIT Press.

[9]   Kohonen, T. (1988), *Self-Organisation and Associative Memory*, Springer-Verlag.

[10] Lin, L-J. (1992), "Self-Improving Reactive Agents Based on Reinforcement Learning, Planning and Teaching," *Machine Learning, (8)*, pp.293-321.

[11] Lin, L-J (1993), *Reinforcement Learning for Robots Using Neural Networks*, Ph.D. Thesis, School of Computer Science, Carnegie Mellon University, Pittsburg, USA.

[12] Maes, P. and Brooks, R. (1990), "Learning to Coordinate Behavior," *Proceedings of the Eighth National Conference on Artificial Intelligence*, Morgan Kaufmann, San Mateo, CA, pp. 796-802.

[13] Mahadevan, S. and Connell, J. (1991), "Scaling Reinforcement Learning to Robotics by Exploiting the Subsumption Architecture," *Proceedings of the Eighth International Workshop on Machine Learning*, Evanston, Illinois, Morgan Kaufmann, pp. 328-332.

[14] Mataric, M.J. (1994), *Interaction and Intelligent Behavior*, Ph.D. Thesis, Massachusetts Institute of Technology.

[15] Millàn, J. and Torras, C. (1992), "A Reinforcement Connectionist Approach to Robot Path Finding in Non-Maze-Like Environments," *Machine Learning*, 8(3/4), pp. 363-395.

[16] Mondada, F., Franzi, E., and Ienne, P. (1993), "Mobile Robot Miniaturisation : A Tool for Investigation in Control Algorithms," *Proceedings .of the Third International Symposium on Experimental Robotics*, Kyoto, Japan.

[17] Ritter, H.J., Martinetz, M., and Schulten, K.J. (1989), "Topology Conserving Maps for Learning Visuo-Motor Coordination," *Neural Networks*, 2, pp.159-168.

[18] Ritter, H.J., Martinetz, M., and Schulten, K.J. (1992), *Neural Computation and Self-organizing Maps*, Addison-Wesley Publishing Company.

[19] Rumelhart, D.E., Hinton, G.E., and Williams, R.J. (1986), "Learning Internal Representations by Error Propagation," *Parallel Distributed Processing*, Cambridge, Massachusetts, MIT Press Edition, Vol 1, Chapter 8.

[20] Sourouchyari, M. (1989), "Mobile Robot Navigation: A neural network approach," *Group CARNAC edition*, EPFL-UNIL Lausanne.

[21] Sehad, S. and Touzet, C. (1994), "Self-organizing Map for Reinforcement Learning : Obstacle-Avoidance with Khepera," *From Perception To Action, PERAC'94*, Lausanne Suisse, September, IEEE Computer Society Press.

[22] Sehad, S. and Touzet, C. (1995), "Neural Reinforcement Path Planning for the Miniature Robot Khepera," *Proceedings of the World Congress on Neural Networks,* Vol 2, pp.350-354, Washington D.C., USA, INNS Press.

[23] Sutton, R. S. (1984), *Temporel Credit Assignment in Reinforcement Learning*, Ph.D. Thesis, Department of Computer and Information Science, University of Massachusetts.

[24] Sutton, R.S. (1990), "Integrated Architectures for Learning, Planning, and Reacting based on Approximating Dynamic Programming," *Proceedings of the Seventh International Conference on Machine Learning*, pp. 216-224.

[25] Sutton, R.S. (1992), "Reinforcement Learning Architectures for Animats," *Proceedings of the First International Conference on Simulation of Adaptive Behaviour, From Animals to Animats*, Edited by J-A Meyer and S. W. Wilson, pp. 288-296.

[26] Watkins, C.J.C.H. (1989), *Learning from Delayed Rewards*, Ph.D. Thesis, King's College, Cambridge.

[27] Watkins, C.J.C.H. and Dayan, P. (1992), "Technical Note: Q-learning," *Machine Learning*, (8), pp. 279-292.

Chapter 8

NEURAL FUZZY TECHNIQUES IN SONAR-BASED COLLISION AVOIDANCE

I. Ahrns *
Research & Technology
Daimler-Benz AG
Ulm, Germany

J. Bruske
Computer Science Institute
Christian-Albrechts-
University
Kiel, Germany

G. Hailu
Computer Science Institute
Christian-Albrechts-
University
Kiel, Germany

G. Sommer
Computer Science Institute
Christian-Albrechts-
University
Kiel, Germany

In this chapter we report application of neuro-fuzzy control to sonar based collision avoidance of our TRC labmate robot, Figure 5. To this end, we will first provide the reader with a brief overview of existing concepts of neuro-fuzzy control and then present our own approach based on Radial Basis Functions. This particular Fuzzy-RBF (FRBF) approach is innovative w.r.t. three aspects of neuro-fuzzy control. First, it alleviates the covering problem in fuzzy control, i.e. the problem of an exponential growth of the number of rules with the dimension of the input space. Second, it provides a means for exact interpolation, i.e. inspite of overlapping membership functions the output of the controller can be guaranteed to take the value of the i-th rule if it has degree of fulfillment one. Finally, by using DCS, [1], instead of RBF networks, output calculation of the controller is very fast on average, since only a few rules (the best matching ones) are evaluated on presentation of an input to the controller.

Utilizing FRBF-based controllers we then present two solutions to the collision avoidance problem faced by mobile robots. The first one is a reactive, behavior-based approach in which collision avoidance is implemented as an in-

---

*The work reported in this chapter was performed at Christian-Albrechts-University at Kiel, Germany

dividual high priority behavior. In such an architecture the behavior selection problem must be solved, i.e. in what situations which behavior has to take over control. Consistent with the fuzzy approach we use a fuzzy blending scheme based on a "pain" function. In our second approach to sonar-based collision avoidance we avoid typical deadlock problems by a closer interaction between higher level tasks and FRBF-based collision avoidance as well as by more advanced feature extraction. Here the task of the controller is to safely follow the freespace direction.

A chapter on neuro fuzzy control would not be complete without a demonstration of its learning capabilities. These are illustrated in Section 4, taking reinforcement learning of collision avoidance as an example.

Since reactive approaches to collision avoidance based on sonar sensors only work if highly erroneous readings caused by crosstalk, bad reflection properties of the environment and shielding problems of the sensors can be eliminated, we finally present a sensor pre-processing method based on sensor grouping and a modified extended Kalman filter algorithm. This easy-to-implement method works very well in practice and is compatible with most existing approaches to fuzzy collision avoidance.

1   Neuro-Fuzzy Control

According to a classification put forward in [2] there are two principal approaches to neuro-fuzzy control. The first one is *cooperative neuro-fuzzy control*, in which the fuzzy controller and the neural network remain seperated, the second one is *hybrid neuro-fuzzy control*, in which the fuzzy controller is realized as a neural network. In cooperative neuro-fuzzy control as e.g. employed in [3] the neural network is used for off-line generation of the membership functions or linguistic rules from training data, typically by clustering the data, or the network is used for online adaptation of membership functions or the weighting of rules in the fuzzy controller.

In this chapter emphasis is on hybrid neuro-fuzzy control. In particular, we exploit the functional equivalence between a restricted class of Sugeno-type fuzzy controllers and Radial Basis Function (RBF) networks as observed in [4] and explained in Section 1.1. Utilizing this equivalence, prior knowledge in form of fuzzy rules can be used to prestructure and initialize an RBF network. The latter can then be trained and refined on training data, and the result of training can reversly be interpreted as fuzzy rules.

While theoretically simple, RBF-based hybrid neuro-fuzzy control has a number of practical problems. These problems are alleviated by extending RBF netwoks in a way described in Section 1.2, resulting in Fuzzy RBF networks (FRBF). The applicability of this approach is demonstrated in Section 2 and 3, and its capability of learning in Section 4.

1.1 On the Equivalence between RBF Networks and Sugeno Type Fuzzy Control

Normalized RBF networks are function approximators ($\mathbb{R}^n \to \mathbb{R}^m$) and calculate their output according to the evaluation function

$$y(x) = \frac{\sum_{i=1}^{N} o_i h_i(x)}{\sum_{i=1}^{N} h_i(x)}, \tag{1}$$

where $h_i(x) = \exp(-\frac{(x-c_i)^2}{\sigma^2})$ denotes a Gaussian radial basis function with center $c_i \in \mathbb{R}^n$. We will refer to the $o_i \in \mathbb{R}^m$ as *output vectors*.

Sugeno-type fuzzy controllers, [2], consist of a set of N linguistic fuzzy rules

$$R_i : \textbf{if} \underbrace{\underbrace{X_1 \text{ is } \mu_{i1} \wedge \ldots \wedge X_n \text{ is } \mu_{in}}_{\text{proposition}} \textbf{ then } \underbrace{f_i(x)}_{\text{consequent}}}_{\text{antecedent}}, \tag{2}$$

where $f_i : \mathbb{R}^n \to \mathbb{R}^m$ are functions in x. The output of a Sugeno-type fuzzy controller is computed according to the defuzzyfication formula

$$y(x) = \frac{\sum_{i=1}^{N} f_i(x)\tau_i(x)}{\sum_{i=1}^{N} \tau_i(x)}, \tag{3}$$

where $\tau_i(x)$ is the degree of fulfillment of the i-th antecedent. Now (1) and (3) become identical if

- consequent functions f_i are restricted to constant functions, i.e $f_i(x) = o_i$,

- membership functions $m_{lj}(x)$ are restricted to gaussians, i.e. $m_{lj}(x_l) = \exp(-\frac{(x_l - \mu_{lj})^2}{\sigma^2})$, where $T(X_l) = \{T_{l1}, \ldots, T_{lk_l}\}$ is the linguistic term set of the linguistic variable X_l, ($1 \leq l \leq N$) with membership functions $M(x_l) = \{m_{l1} \ldots, m_{lk_l}\}$ and

- the fuzzy conjunction is implemented as the algebraic product.

In this case, (3) can be written as

$$y(x) = \frac{\sum_{i=1}^{N} o_i \exp(-\frac{(x-c_i)^2}{\sigma^2})}{\sum_{i=1}^{N} \exp(-\frac{(x-c_i)^2}{\sigma^2})} = \frac{\sum_{i=1}^{N} o_i h_i(x)}{\sum_{i=1}^{N} h_i(x)}, \tag{4}$$

with $c_{ij} = \mu_{ij}$ for proposition "X_j is μ_{ij}" in the antecedent of the i-th rule.

1.2 The Fuzzy-RBF Network

Problems with RBF-based neuro-fuzzy control are that the number of nodes (fuzzy rules) grows exponentially with the dimension of the input space and that due to overlap at the centers the basis (membership) functions interfere with each other. Another problem is that all rules need to be evaluated, even if they have a very low degree of fulfillment. In the following, we show how exponential growth of the number of rules can be circumvented by incompletely specified antecedents. We also provide a solution to the inference problem. Finally, we show how Dynamic Cell Structures (DCS), [1], help to avoid evaluating all rules by evaluating only the best matching rule and its topologically neighbors.

1.2.1 Incomplete Rules

In high dimensional input spaces an RBF-based controller faces the problem of an exponential explosion of the number of nodes (rules) if the input space has to be uniformly covered. If the input dimension is n and we have a rule depending on only k variables, this rule has to be expanded to l^{n-k} rules if each variable takes l linguistic values. This problem can be alleviated if we drop the requirement that each node in the network must compute an activation function in n dimensions. Instead we must allow for nodes computing an activation function in just k dimensions, if k is the number of input variables in the linguistic rule. For that purpose, we introduce the additional symbol $\perp$, denoting an undefined value, and set c_{ik}, the k-th component of the i-th center to

$$c_{ik} = \begin{cases} \mu_{ik} & : \quad \text{proposition } X_k \text{ is } \mu \text{ belongs to the antecedent of the } i\text{-th rule} \\ \perp & : \quad X_k \text{ does not appear in the antecedent of the } i\text{-th rule} \end{cases} \tag{5}$$

The activation function corresponding to that antecedent is calculated as

$$h_i(x) = \exp(- \sum_{k \in \{1,\ldots,N\} \setminus \{l|c_{il}=\perp\}} \frac{(x_k - c_{ik}^2)}{\sigma^2}). \tag{6}$$

A similar problem arises for MISO systems with an m dimensional output and rules which only effect l of these values. Again, with $\perp$ denoting an undefined value, we set o_{ik}, the k-th component of the i-th output vector to

$$o_{ik} = \begin{cases} f_{ik} & : \quad \text{k-th output component in consequent of the } i\text{-th rule} \\ \perp & : \quad \text{k-th output component not effected by } i\text{-th rule} \end{cases} \tag{7}$$

The k-th output of the FRBF controller is then obtained as

$$y_k = \frac{\sum_{i \in \{1,\ldots,N\} \setminus \{l|o_{ik}=\perp\}} o_{ik} h_i(x)}{\sum_{i \in \{1,\ldots,N\} \setminus \{l|o_{ik}=\perp\}} h_i(x)}, \tag{8}$$

where the activation function $h_i(x)$ is calculated according to (6).

1.2.2 Exact Interpolation

Another problem with RBF networks used for Sugeno-type fuzzy control is that the networks do not exactly interpolate between the output vectors (consequent values of the rules) but rather perform an approximation. This is due to the overlap of the basis functions at their centers: Even if one activation function (degree of fullfilment of an antecedent) takes the value one, the output of the controller deviates from the corresponding consequent value. However, exact interpolation sometimes is necessary if the control function must fulfill certain constraints. In obstacle avoidance, for instance, this could be a rule which sets the forward velocity to zero if the frontal distance reaches a certain value. This important rule must not be biased by other rules and probably will not be subjected to adaptation as well. Our solution to the exact interpolation problem is to introduce a set Π of *exact nodes* and to modify the activation functions $h_i(x)$ by multiplying the original activation functions with the complement of each activation function of nodes in Π, i.e.

$$h'_i(x) = h_i(x) \prod_{j \in \Pi \setminus i} (1 - h_j(x)), \tag{9}$$

which assures that if one membership function has degree of fulfillment one, the others will be suppressed and hence do not contribute to the mapping. For exact nodes we require that they have no undefined components in their output vectors and centers. A formal proof of the exact interpolation property can be found in [5].

The complete FRBF architecture is illustrated in Figure 1.

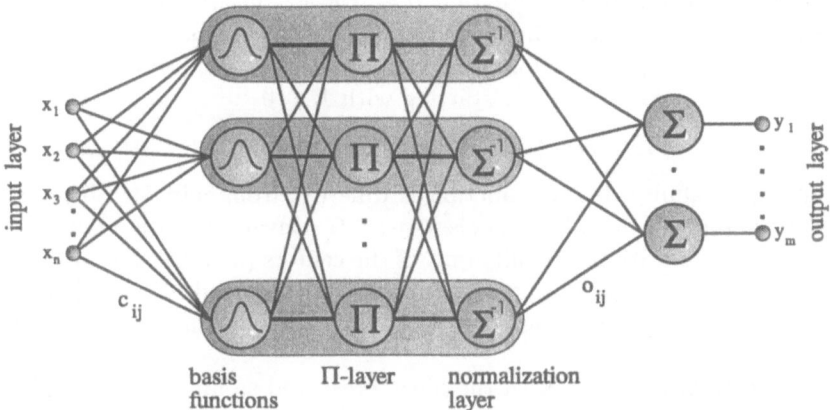

Figure 1: FRBF architecture with exact nodes: Each formal node in the nework consists of a basis function, a π neuron for exact interpolation and a normalization unit, which may be grouped in three computational layers.

1.2.3 Topology Preservation

DCS networks represent an extension of RBF-networks in that each node in the network gets to know (learns) which other nodes in the network are its nearest neighbors in input space. The rationale for this is that activation of the nodes decreases with decreasing distance of the input from their centers, and hence only the best matching unit bmu (i.e. the node whose center c_i has smallest Euclidean distance to the input value x) and its direct neighbors significantly contribute to the mapping and need to be evaluated. But how can direct neighborhood be defined? The intuitive definition was provided by T. Martinetz, [6], calling two nodes neighbored in input space if their masked Voronoi cells[1] have a common border. This leads to the definition of the *induced Delaunay triangulation* or *perfectly topology preserving map* in which two nodes are connected if their masked Voronoi cells share a common border. If $G = (V, E)$, $V = \{1, \ldots, N\}$ is the graph representing the induced Delaunay triangulation of the centers $c_1, \ldots, c_N$ of the network, then the set of direct neighbors of node i, $Nh(i)$, can be defined as the direct neighbors w.r.t. G, i.e.

$$Nh(i) = \{j | (i, j) \in E\}. \tag{10}$$

The output of the network is calculated as

$$y(x) = \frac{\sum_{i \in Nh(bmu)} o_i h_i(x)}{\sum_{i \in Nh(bmu)} h_i(x)} \tag{11}$$

and includes only the best matching nodes (rules). This can lead to a significant speed up in output calculation.

The induced Delaunay triangulation (i.e. the graph G) for a given set of centers can be either calculated in advance or can be approximated with a simple learning rule that, on presentation of an input x, always connects the best and the second best matching unit, i.e. starting with $E = \emptyset$

$$E = E \cup \{(bmu, smu), (smu, bmu)\}. \tag{12}$$

If the input probability density function is different from zero for all possible inputs and the distribution of centers is *dense*[2], G converges to the induced Delaunay triangulation with probability one. If the centers (membership functions) are allowed to adapt, existing edges may need to be removed from G. This can be achieved by decaying existing edges by a learning rule such as proposed in [7].

Combining DCS networks with the ideas presented so far (incomplete rules and exact interpolation) we obtain the Fuzzy DCS (FDCS). Yet there remains a problem within the FDCS framework: How can a best matching unit be determined (and the topology be defined and learned) if components of the centers may be undefined? The intuitive answer is to replace the Euclidean metric with the distance measure

$$d(x, c_i) = \sqrt{\sum_{k \in \{1, \ldots, n\} \setminus \{l | c_{il} = \perp\}} (x_k - c_{ik}^2)}, \tag{13}$$

---

[1] If $M \subset \mathbb{R}^n$ is the input manifold and $V_i \subset \mathbb{R}^n$ the Voronoi cell of c_i then the maksed Voronoi cell is $V_i^{(M)} = V_i \cap M$.

[2] The distribution of centers is dense if for each possible input $x \in M$ the triangel formed by x, c_{bmu} and c_{smu} completely lies in M.

which takes into account only the defined components. With help of this distance measure, the notions of topology preservation and the learning rules for the topology preserving lateral connection structure as discussed in 1.2.3 can be generalized for the new situation of undefined components, as has been formally proven in [5].

2 Behavior-Based FRBF Control for Collision Avoidance

After this short introduction into our FRBF-based approach to fuzzy control we will now propose a fully reactive and behavior-based control architecture for obstacle avoidance.

The main problem with behavior-based control is the switching between different behaviors and that in order to exploit the advantages of smooth fuzzy control this switching additionally has to be smooth. We solve this problem by introducing a *pain* function which smoothly switches between the avoidance behavior and the task-driven behavior. The task-driven behavior consists of two subtasks, wall-following in an exploration phase or goal-following to reach a goal position.

2.1 FRBF-Based Architecture for Reactive Collision Avoidance

In the behavior-based subsumption-control-architecture, e. g. [8], the behaviors are organized horizontally, i.e. each behavior has full access to all sensor readings and proposes its own motor control command. The final motor control command is then computed by suppression: Low level behaviors are more important for the safety of the system and therefore have a higher priority than higher level tasks.

Classical subsumption architectures rest on a hard selection of one of their behaviors. Hence these architectures result in a kind of *bang-bang* control on the basis of potentially smoothly controlled *bang*-functions.

In fuzzy control we seek for a smoothly controlled decision between the authorization of two fuzzy controllers. To this end, let us consider two FRBF networks $\mathcal{A}$ and $\mathcal{B}$ which represent two hierarchically ordered behaviors. Here, $\mathcal{A}$ denotes the lower level behavior, more important for the safety of the system, and $\mathcal{B}$ denotes the more task oriented higher level behavior. We write $y_{\mathcal{A}} : I_{\mathcal{A}} \to O$ and $y_{\mathcal{B}} : I_{\mathcal{B}} \to O$ for the corresponding evaluation functions of both networks. The input spaces might even be different, so that the behaviors can be optimally adapted to their special tasks.

For each behavior we now introduce a *state-evaluation* function,

$$eval : I_{\mathcal{A}} \to [a, b], \ a, b \in \mathbb{R}, \tag{14}$$

which signals the danger of the current system state from the point of view of each individual beahvior. The state-evaluation can be regarded as signalling the responsibility of behavior $\mathcal{A}$.

In biological systems we also have a kind of state-evaluation which can be regarded as a pain signal. Pain is able to change the behavior of any living being, if only the pain intensity exceeds some threshold value. Accordingly, we introduce a *pain signal*, $p : [a, b] \rightarrow [0, 1]$, whose response activates the behavior.

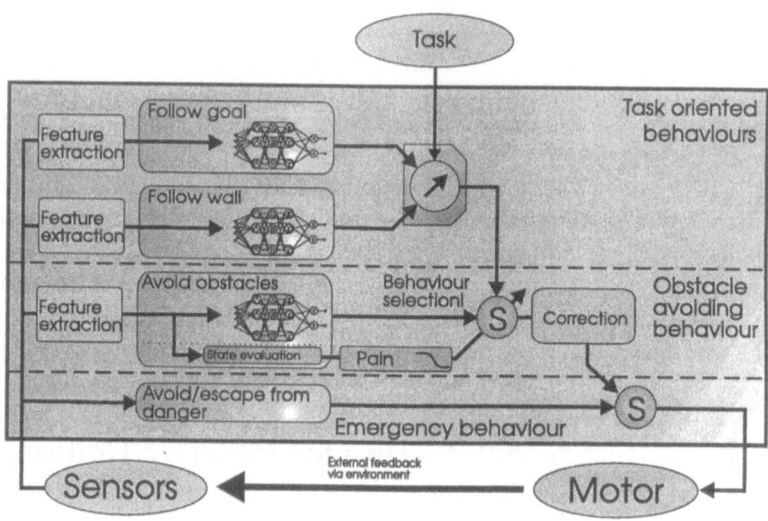

Figure 2: Behavior-based fuzzy architecture for obstacle avoidance

We demand p to be a monotonically decreasing function fulfilling $p(a) = 1$ and $p(b) = 0$. Using the convex-combination

$$y = p(eval(x_A) \cdot y_A(x_A) + (1 - p(eval(x_A)) \cdot y_B(x_B), \qquad (15)$$

we obtain a motor control command $y \in O$ for any input $x_A \in I_A$ and $x_B \in I_B$. For a low state evaluation $eval(x_A)$ the behavior $\mathcal{A}$ dominates, while for high state evaluations the resulting reaction will be dominated by the higher level control network. The values a and b represent the maximum and minimum state evaluation. As an example of a pain signal, consider

$$p(z) \quad = \quad \varphi(z) - \varphi(b) + \frac{\varphi(a) - \varphi(b) - 1}{b - a} \cdot (z - b), \text{ with} \qquad (16)$$

$$\varphi(x) \quad = \quad \frac{1}{1 + \exp(\varrho \cdot (x - m))}. \qquad (17)$$

The parameter $m \in [a, b]$ denotes the location of the reversal point, and $\rho \in \mathbb{R}$ a parameter that stands for the inclination of the pain signal. For $\rho = 0$, we obtain a simple linear function, and for $\rho \rightarrow \infty$ we get a threshold function with $p(z) = 1$ for $z < m$ and $p(z) = 0$ for $z > m$. Figure 3 illustrates the pain signal.

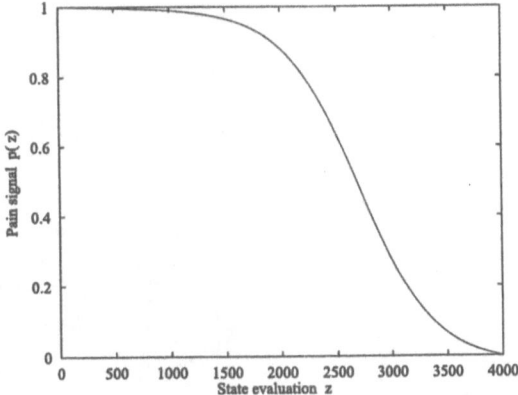

Figure 3: Pain signal for the parameter set $a = 0$, $b = 4000$ and $\varrho = 0.003$

Using a pain signal, we can smoothly combine two different behaviors. Since these behaviors themselve consist of smooth FRBF controllers, we finally obtain a smoothly controlled behavior-based reactive control architecture.

For reasons of safety, however, we added a conventional hard-wired lowest-level emergency behavior that is able to avoid a direct imminent collision by applying an emergency stop and triggering an orientation behavior. Yet all higher level tasks, such as obstacle avoidance, wall following and goal following, result from a smooth combination of different fuzzy controllers.

2.1.1 Feature Extraction

Each FRBF-controller has its own pre-processing stage for feature extraction. During this stage from all available sensory inputs only those features are selected that are relevant for the specific task of each behavior. Such a pre-processing stage is necessary to reduce the dimensionality of the input spaces of the FRBF-controllers. In our experiments, the sensory input consists of all eight sonar readings, the actual translational velocity and the angular velocity (jog-rate). Not all these inputs are necessary for all behaviors.

For the FRBF-controller of the obstacle avoidance behavior we choose as features all eight sonar readings, whereas the features of wall following are just the orientation to the left and to the right wall.

Using the ultrasonic sensor arrangement as shown in Figure 4, we define these two orientation features by

$$\alpha_{right} = \arctan\left(\frac{sensor[5] - sensor[4]}{d_v}\right) \cdot \frac{180°}{\pi}, \tag{18}$$

$$\alpha_{left} = \arctan\left(\frac{sensor[7] - sensor[6]}{d_v}\right) \cdot \frac{180°}{\pi}, \tag{19}$$

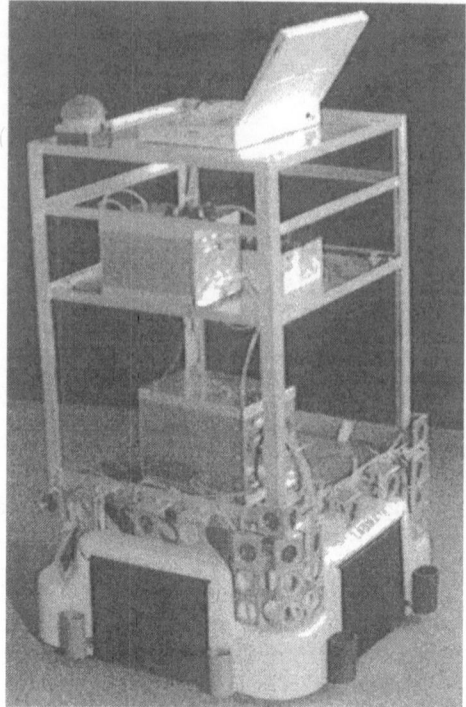

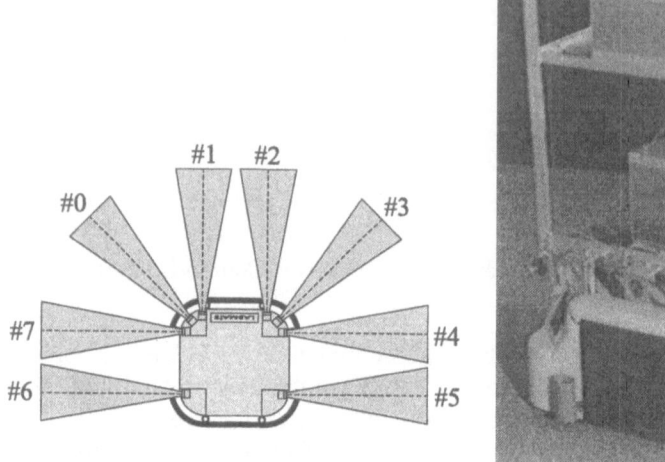

Figure 4: Arrangement of ultrasonic
sensors for the TRC Labmate

Figure 5: The physical robot

where d_v defines the vertical distance of the lateral sensors.

Finally, the feature extraction for the goal following behavior is canonically
defined by the horizontal and vertical components Δx and Δy of the Euclidean
distance between the center of the robot and the goal position in egocentric
robot coordinates.

In addition to the features proposed here, additional features can be utilized.
For example, similar to the inclination to lateral obstacles, the inclination to
frontal obstacles can be defined as a possible feature. If we want to formulate
fuzzy rules in which two sonar readings are compared (e.g. $s_1 > s_2$), we introduce
the difference $(s_1 - s_2)$ as an additional feature.

2.1.2 Correction of Direct Motor Commands

An additional safety component checks whether the motor commands are within
a tolerable interval or not. In particular, such post-processing is recommended
if the FRBF controller is allowed to adapt, cf. Section 4. In order to make sure
that only reasonable outputs reach the motors, we filter the proposed commands
by

$$\tilde{a} = \begin{cases} a & : & a \in [l_a, r_a] \\ l_a & : & a < l_a \\ r_a & : & a > r_a \end{cases} . \qquad (20)$$

Filtering the proposed value $a \in \{\omega, v\}$ finally yields a corrected value $\tilde{a}$ which lies in the admissable intervall $[l_a, r_a]$. The value v denotes the velocity and ω the superimposed angular velocity.

A simple correction mechanism that guarantees a driving velocity with which the robot cannot contact an obstacle in the distance d_{stop} before T seconds is obtained by choosing

$$r_v = \frac{d_{min} - d_{stop}}{T}, \qquad (21)$$

where d_{min} is the minimum frontal distance to any obstacle.

2.2 Fuzzy Rules for the FRBF Controller

In this section we provide the fuzzy rule base for the FRBF controller. Instead of linguistic terms we directly give numerical values. The values of the premises make up the centers, as described in Section 1.1. The sign $\approx$ will be used whenever conventional (inexact) neural units are used. Correspondingly, the use of exact neural units is denoted by the $=$ sign. Sonar readings are denoted by s_i.

2.2.1 Collision Avoidance Behaviors

Obstacle avoidance is often divided into sub-behaviors that treat special types of obstacle situations. For instance, in [8] one distinguishes between sub-tasks as *avoiding frontal obstacles* or *avoiding lateral obstacles*.

Our fuzzy controller has no inherent structure or hierarchy of rules, yet we arrange the rules in different groups of neural units that can be interpreted as *velocity control units, frontal collision avoiding units* and *lateral collision avoiding units*. The fuzzy rules are shown in Tables 1, 2 and 3. Figure 6 illustrates the evaluation function of the resulting FRBF controller.

---

if $(s_1 \approx 250)$ **then** $v \approx 0$
if $(s_2 \approx 250)$ **then** $v \approx 0$

if $(s_1 \approx 2000) \wedge (s_2 \approx 2000) \wedge (s_0 \approx 500) \wedge (s_3 \approx 500)$ **then** $v \approx 500$
if $(s_1 \approx 2000) \wedge (s_2 \approx 2000) \wedge (s_0 \approx 1000) \wedge (s_3 \approx 1000)$ **then** $v \approx 600$
if $(s_1 \approx 2000) \wedge (s_2 \approx 2000) \wedge (s_0 \approx 2000) \wedge (s_3 \approx 2000)$ **then** $(v = 1000) \wedge (\omega = 0)$

if $(s_1 \approx 1500) \wedge (s_2 \approx 1500)$ **then** $v \approx 400$

if $(s_1 \approx 1000) \wedge (s_2 \approx 1000)$ **then** $v \approx 300$

if $(s_1 \approx 500) \wedge (s_2 \approx 500)$ **then** $v \approx 150$

Table 1: Rules for controlling driving velocity

ω	s_0				
s_3	0	300	700	1200	1800
0	0	30	30	30	30
300	-30	0	30	30	30
700	-30	-30	0	30	30
1200	-30	-30	-30	0	30
1800	-30	-30	-30	-30	0
ω	s_1				
s_2	0	300	700	1200	1800
0	0	30	30	30	30
300	-30	0	30	30	30
700	-30	-30	0	30	30
1200	-30	-30	-30	0	30
1800	-30	-30	-30	-30	0

Table 2: Rules for avoiding frontal obstacles

ω	s_4			
s_5	0	300	700	1200
0	30			
300	30	30		
700	30	30	30	
1200	30	30	30	30
ω	s_7			
s_6	0	300	700	1200
0	-30			
300	-30	-30		
700	-30	-30	-30	
1200	-30	-30	-30	-30

Table 3: Rules for avoiding lateral obstacles

2.2.2 Task Oriented Behaviors

Table 4 shows the rule base for the wall following and the goal-following behavior. These rules are realized by exact neural units. Only by using exact neural units we can assure that the driving velocity of the vehicle exactly becomes zero at the goal position. Due to superposition of several units this would hardly be possible with inexact neural units.

2.3 Simulations

In our simulations, the FRBF-based approach for obstacle avoidance turned out to be superior to the classic behavior-based approaches, e. g. [8], we investigated for comparison. The FRBF-based approach was able to achieve a higher maximum speed when there were only far obstacles. In contrast to the maximum speed of $0.25 \frac{m}{s}$ achieved by the classic approach, the fuzzy controller

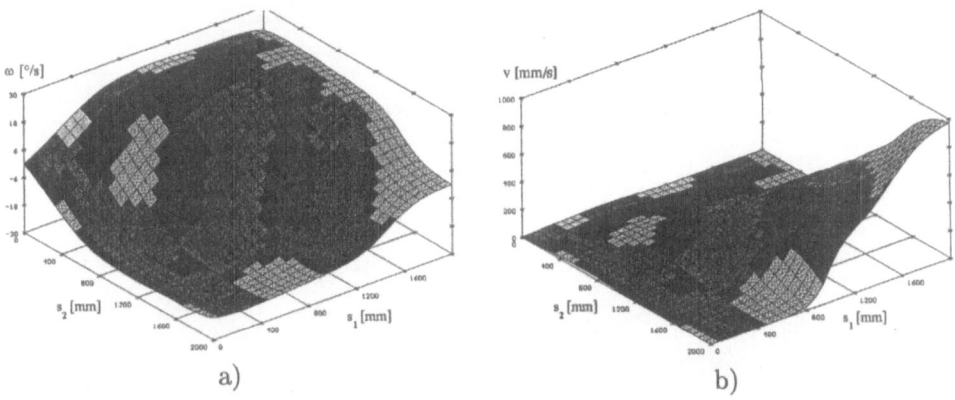

Figure 6: Evaluation function of the FRBF controller for fixed input values $s_0 = s_3 = \cdots = s_7 = 2000\ mm$.

Goal following			
v/ω	Δx		
Δy	-2000	0	2000
-2000	0/-30	0/30	0/30
0	0/-30	0/0	0/30
2000	400/-30	800/0	400/30

Wall following			
v/ω	α_{left}		
α_{right}	-50	0	50
-50	800/-30	800/-30	
0	800/-30	800/0	800/30
50		800/30	800/30

Table 4: Fuzzy rules for wall- and goal following behavior

nearly reached the maximum speed of $1\frac{m}{s}$. Nevertheless it was careful in narrow environments where it navigated with very low speed. In extremely narrow environments, the FRBF network controls the forward velocity down to zero and a remaining rotational component results in an "escape from danger" behavior.

Furthermore the smooth control by the FRBF controller was observed to produce fewer oscillations during a corridor passage. The wall following behavior is able to align the vehicle with walls. Figure 7 shows the robot in a simulated environment. The velocity can be recognized from the distance of the dashes perpendicular to the driving direction. Higher velocities produce more distant dashes while lower velocities produce narrower dashes. In environments with far obstacles the vehicle reaches a velocity of nearly $0.8\frac{m}{s}$.

For all FRBF controller we used Gaussian neural activation functions with a standard deviation of $\sigma = 0.2$. Inputs to the network were normalized. As a state evaluation we defined

$$eval(s_0, \cdots, s_7) = \sqrt{s_0^2 + s_1^2 + s_2^2 + s_3^2}. \tag{22}$$

The maximum measurable distance (timeout distance) of the ultrasonic sensors was set to $2m$, so the parameter a and b were set to $a = 0$ and $b = 4000$. For the pain signal we have chosen $m = 2700$ and $\varrho = 0.003$. The time step width was

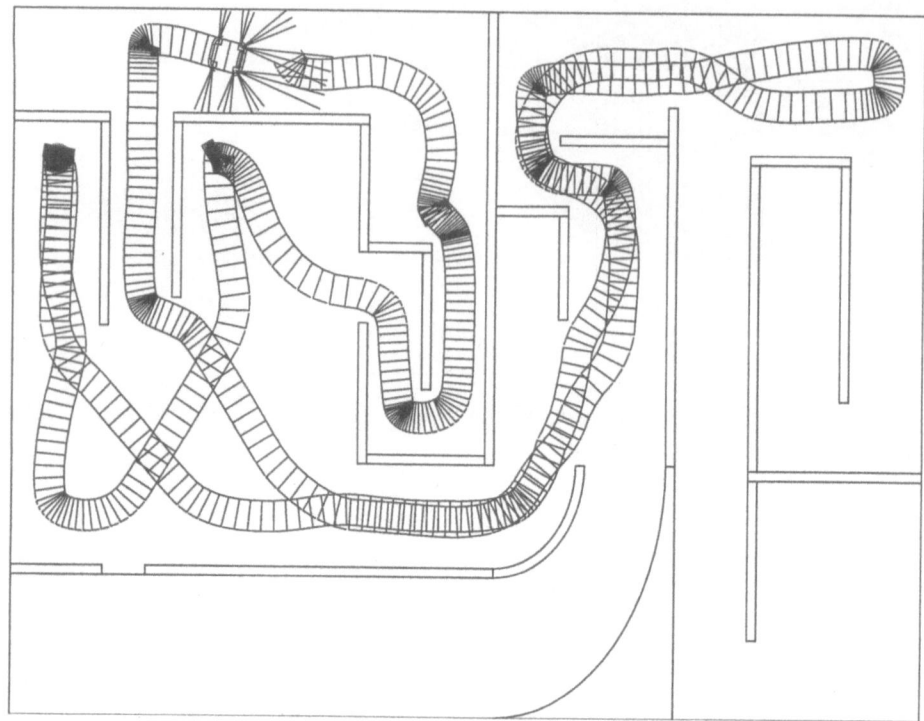

Figure 7: FRBF controlled tour of a simulated autonomous mobile robot

set to $0.4s$ and corresponds to the time interval for one control cycle in reality.

There is, however, a problem with the proposed control scheme which has to be addressed. In contrast to the classical subsumption architecture where each behavior proposes its own motor command and finally only one command is selected by a hierarchy of suppression there is no possibility of prefering specific rules in the FRBF network architecture. All neural units contribute their activation-driven output to the motor command, which is finally send to the motor. For example, in the classical subsumption architecture testing whether left rotation is an appropriate reaction before testing for right rotation results in a behavior which tends to prefers left curves . Such behavior can be regarded as *non-symmetric*. On the contrary, the FRBF architecture has no such built in preference and will typically result in a *symmetric* behavior (cf. Figure 6).

Symmetric controllers cause an undesirable behavior (in simulations as well as in reality) when the robot is driving into a corner. For simplicity we only consider the sonar readings # 1 and # 2. At first, both readings decrease and as a result the vehicle decreases its velocity. If the left sensor shows a smaller reading than the right sensor, the vehicle will turn to the right and vice versa. But now, by turning towards one direction, because of the new environmental situation the state of the sonar readings will switch and the symmetric controller will turn to just the opposite direction in the next step. This behavior results in a kind of *dead lock situation*, shown in Figure 8, which normally should be avoided. Until now, only the hard wired emergency behavior can free the robot from the dead lock situation.

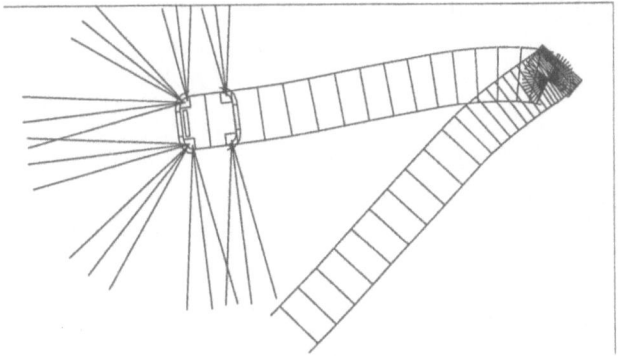

Figure 8: A dead lock situation

3   Freespace-Based FRBF Control for Collision Avoidance

To overcome the dead lock problem mentioned above and keep on profiting from the advantages of fuzzy control, we next propose a FRBF based approach that uses improved feature extraction from sonar readings and only consists of a single FRBF network.

3.1   Feature Extraction

In order to avoid a symmetric fuzzy controller we change the input features of the controller and introduce an angle that points into the direction of a free space that is closest to the heading of the vehicle. This free space direction provides an additional input feature and can be obtained by a technique which is similar to the obstacle avoidance approach of [9].

In [9], the last three sonar readings are stored and build a coarse temporally and spatially restricted model of the environment. From this model an optimal trajectory, chosen from a set of circles through the center of the robot and tangential to the heading direction, is generated and provides the obstacle avoiding behavior.

What we did is to extract the free space direction from the last stored sensor readings and provide it as an additional input feature for our fuzzy controller.

Because no reliable free space information can be estimated from only eight ultrasonic sensor readings, we also store the sonar readings in a short time memory. From this short time memory we extract a coarse model of the direct robot surrounding. In contrast to [9] we do not use this model to explicitly estimate a trajectory, but we recover the free space direction closest to the actual heading direction. Figure 9 shows the short time representation of the environment from which a free space direction is generated.

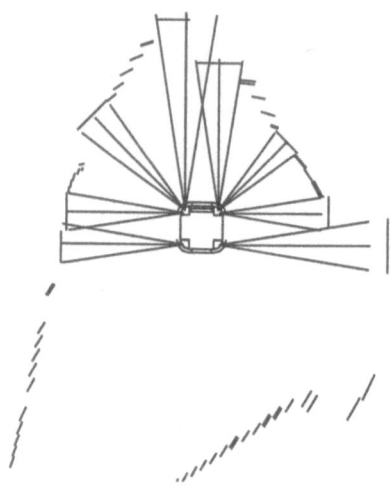

Figure 9: Short time representation of the robot surrounding

From the short time memory, additional features can be extracted, e. g.

- Direction of the free space closest to the goal direction. This will result in a goal following behavior.

- Direction of the free space closest to the actual robot heading. This will result in a wandering behavior.

- Distance to frontal or lateral obstacles.

- Orientation of frontal or lateral walls.

3.2   FRBF-Based Architecture for Collision Avoidance with Freespace Information

In this section we investigate FRBF control based on free space extraction for obstacle avoidance and show that it can solve the dead lock problem of the behavior-based approach.

Eight sonar readings and the supplementary extracted free space direction make up the new input features. As in the previous section two higher level tasks are considered, namely wall alignment (in order to provide an exploration behavior) and goal following. So far these two tasks have been divided into two different control modules. Using the additional free space information, this division is no longer necessary. Both tasks can be solved by choosing the free space direction in accordance with the particular task.

Goal following behavior: If the robot has to reach a specified position, the free space direction is selected that is closest to the direction of the goal position.

Wandering around behavior: Exploration of the robot surrounding can be achieved by a free space direction that is closest to the robot heading. If the robot moves towards a wall, the free space direction will align to the wall direction. Therefore, following the free space direction produces a wall following behavior.

The resulting controller architecture consists of a single FRBF controller together with the correction stage and the additional hard-wired emergency behavior on the lowest level. The new architecture is shown in Figure 10.

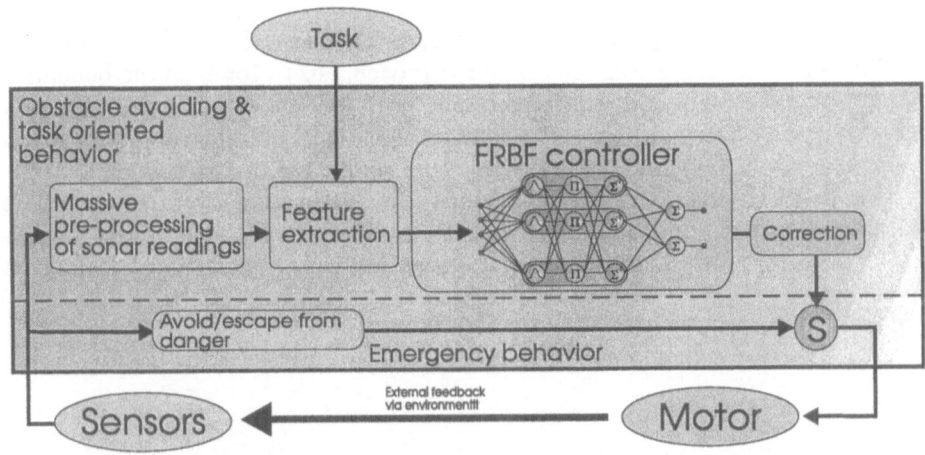

Figure 10: FRBF-based obstacle avoidance using massive sonar reading pre-processing

3.3 Fuzzy Rules of the FRBF Controller

Tables 5-8 declare the fuzzy rules of the FRBF controller used for free space driven obstacle avoidance. As input features we use all eight sonar readings $s_0, \cdots, s_7$, as well as the free space direction α_{best} according to the specified task. Supplementary, we use the rules of Table 3. The fuzzy rules of Table 8 represent exact neural units.

v/ω	α_{best}					
	-150	-100	-50	50	100	150
$s_1 \approx 100:$	0/-30	0/-30	0/-30	0/30	0/30	0/30
$s_2 \approx 100:$	0/-30	0/-30	0/-30	0/30	0/30	0/30

Table 5: Rules dealing with emergency situations

if $(s_1 \approx 3000) \wedge (s_2 \approx 2000) \wedge (s_0 \approx 600) \wedge (s_3 \approx 600)$ then $v \approx 600$

if $(s_1 \approx 3000) \wedge (s_2 \approx 2000) \wedge (s_0 \approx 1300) \wedge (s_3 \approx 1300)$ then $v \approx 750$

if $(s_1 \approx 3000) \wedge (s_2 \approx 3000) \wedge (s_0 \approx 3000) \wedge (s_3 \approx 3000)$ then $(v = 1000) \wedge (\omega = 0)$

if $(s_1 \approx 1500) \wedge (s_2 \approx 1500)$ then $v \approx 400$

if $(s_1 \approx 1000) \wedge (s_2 \approx 1000)$ then $v \approx 300$

if $(s_1 \approx 500) \wedge (s_2 \approx 500)$ then $v \approx 150$

Table 6: Rules for controlling the driving velocity

3.4   Simulations

The experiments we have carried out applying the above FRBF-based architecture demonstrate the workability of our approach, i.e. to combine the benefits of smooth fuzzy control and to overcome the dead lock problem. Remember that we only had to introduce an extra feature which was obtained straightforward by storing the last sensor readings and some simple feature extraction.

Figure 11 b) demonstrates a successful avoiding maneuver in a situation where the previous approach has failed. In Figure 11 a) the extracted free space direction and the sensor readings as stored in the short time memory are depicted.

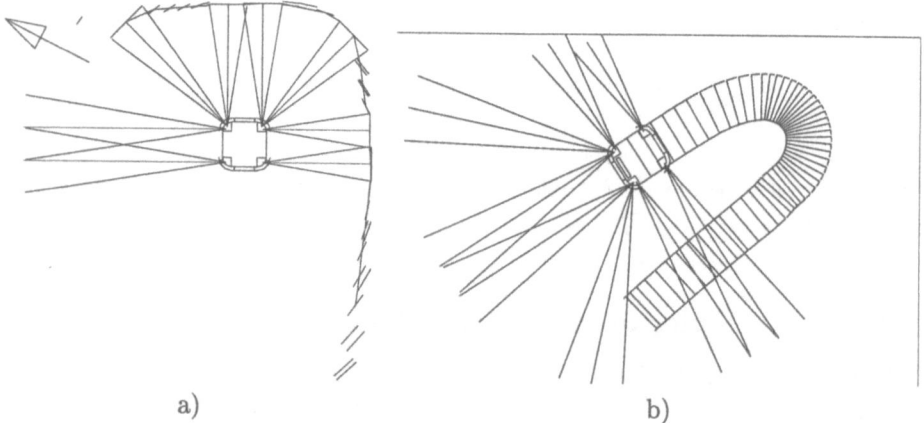

a)   b)

Figure 11: a) Extracted free space direction b) Simulated avoiding maneuver

$\alpha_{best} \approx 0$:					
ω			s_0		
s_3	0	300	700	1200	1800
0		30	30	30	30
300	-30		30	30	30
700	-30	-30		30	30
1200	-30	-30	-30		30
1800	-30	-30	-30	-30	

$\alpha_{best} \approx 0$:					
ω			s_1		
s_2	0	300	700	1200	1800
0		30	30	30	30
300	-30		30	30	30
700	-30	-30		30	30
1200	-30	-30	-30		30
1800	-30	-30	-30	-30	

ω	α_{best}					
$s_0 = s_3$	-150	-100	-50	50	100	150
0	-30	-30	-30	30	30	30
300	-30	-30	-30	30	30	30
700	-30	-30	-30	30	30	30
1200	-30	-30	-30	30	30	30
1800	-30	-30	-30	30	30	30

ω	α_{best}					
$s_1 = s_2$	-150	-100	-50	50	100	150
0	-30	-30	-30	30	30	30
300	-30	-30	-30	30	30	30
700	-30	-30	-30	30	30	30
1200	-30	-30	-30	30	30	30
1800	-30	-30	-30	30	30	30

Table 7: Rules for avoiding frontal collisions

4 Reinforcement Learning of an FDCS Controller for Collision Avoidance

The reason for using *neuro*-fuzzy control is to allow for adaptation. In this section we want to briefly[3] demonstrate the learning capabilities of an FDCS controller by reinforcement learning of collision avoidance. Circumstances are very hard because i) the reinforcement signal provides only minimal feedback from the environment, ii) the reinforcement signal is delayed, iii) we start with only one fuzzy rule and the learner has to generate new rules from interaction

---

[3] A more detailed description of this experiment as well as learning parameters can be found in [10] and [5] .

ω	α_{best}					
	-150	-100	-50	50	100	150
$s_1 \approx s_2 \approx 3000$:	-30	-30	-30	30	30	30

Table 8: Rules for free space following

with the environment, iv) we allow for adaptation of membership and consequent functions and v) we additionally learn the topology of the input space and hence an FDCS controller, cf. Section 1.2.3, for accelerated output calculation. On the one hand, our experiments confirm that even in this extreme setup the controller is indeed able to learn an obstacle avoidance behavior. On the other hand, our experiments also underline that pure reinforcement learning (without prior knowledge and without additional learning mechanisms) takes too much time and too many trial to be of much practical value.

4.1   Learning Architecture and Algorithms

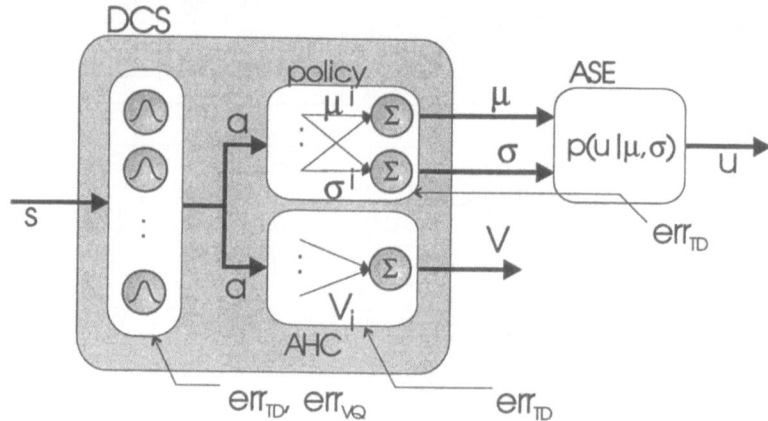

Figure 12: FDCS-based reinforcement learning architecture

As depicted in Figure 12 the controller is realized by a single DCS network with an additional stochastic associative search element (ASE) for REINFORCEment learning. The DCS network implements both the actual controller (policy) as well as an adaptive heuristic critique (AHC), [11]. The calculation of the control vector u given the sensory input s proceeds as follows:

First, the input vector is transformed to an activity vector representing the normalized activations (degrees of fulfillment) of the RBF units (rules) with centers c_i and uniform width:

$$a_i = \frac{h_i(s)}{\sum_{j \in Nh(bmu)} h_j(s)} . \tag{23}$$

Second, a *prototypical action vector* μ, a *certainty vector* σ and the *predicted cumulative reinforcement* V are calculated by a weighted sum of contributing vectors (consequent functions) attached to the RBF units[4]:

$$\mu = \sum_{i \in Nh(bmu) \cup \{bmu\}} a_i \mu^i \ , \ \sigma = \sum_{i \in Nh(bmu) \cup \{bmu\}} a_i \sigma^i \ , \ V = \sum_{i \in Nh(bmu) \cup \{bmu\}} a_i V_i .$$

(24)

Finally the ASE draws the actual action vector u according to a Gaussian probability density function with marginal distributions

$$p(u_j|s) = p(u_j|\mu_j(s), \sigma_j(s)) = \frac{1}{\sqrt{(2\pi)}\sigma_j} e^{-\frac{(u_j - \mu_j)^2}{2\sigma_j^2}}$$

(25)

On-line adaptation is performed w.r.t. the contributing prototypical action vectors μ^i, certainty vectors σ^i and evaluation values V_i attached to the RBF units (consequent - part) as well as to the centers c_i of the RBF units (antecedent - part). The evaluation values V_i are updated using a TD(1) rule [12]:

$$\Delta V_i = \alpha_V \, err_{TD}(s_t) \sum_{k=1}^{t} (\lambda\gamma)^{t-k} \nabla_{V_i} V(s_k)$$

(26)

$$= \alpha_V \, (\gamma V(s_{t+1}) + r_t - V(s_t)) \sum_{k=1}^{t} (\lambda\gamma)^{t-k} a_i(s_k).$$

(27)

where $err_{TD}(s_t) = r_t - b_t$ denotes the current temporal difference error with r_t the reinforcement signal and $b_t = V(s_{t+1}) - V(s_t)$ an adaptive baseline. Prototypical action vectors μ^i and certainty vectors σ^i are adapted using a REINFORCE gradient descent:

$$\Delta \mu_j^i = \alpha_\mu \, err_{TD}(s_t) \frac{\partial \ln p_j(u_j|\mu_j, \sigma_j)}{\partial \mu_j} \frac{\partial \mu_j}{\partial \mu_j^i}$$

(28)

$$= \alpha_\mu \, (\gamma V(s_{t+1}) + r_t - V(s_t)) \frac{(u_j - \mu_j)}{\sigma_j^2} a_i$$

(29)

and

$$\Delta \sigma_j^i = \alpha_\sigma \, err_{TD}(s_t) \frac{\partial \ln p_j(u_j|\mu_j, \sigma_j)}{\partial \sigma_j} \frac{\partial \sigma_j}{\partial \sigma_j^i}$$

(30)

$$= \alpha_\sigma \, (\gamma V(s_{t+1}) + r_t - V(s_t)) \frac{(u_j - \mu_j)^2 - \sigma_j^2}{\sigma_j^3} a_i$$

(31)

The REINFORCE framework [13] states that (28) and (30) implement a gradient descent on the expected reinforcement (at least for a constant baseline b). When the algorithm converges towards a local maximum of the reinforcement the σ_j^i will decrease to small values, narrowing the range of stochastic search. Hence the

---

[4]the contributing prototypical action vectors and certainty vectors are denoted by superscripts.

term certainty values: If we pre-structure the network with fuzzy rules we can specify the search range for the conclusion of this rule by specifying its σ^i vector. Values close to zero result in non-changing consequents (fixed rules). On the other hand, if we analyze the network at consecutive time steps, non decreasing components of σ^i_j indicate convergence to (certainty about) the corresponding prototypical action.

Finally, the centers c_i of the *bmu* and its topological neighbors are updated according to an error modulated Kohonen rule as described in [7]:

$$\Delta c_i = \epsilon_{modulated} \, err_{VQ} \quad \text{with} \quad err_{VQ} = s - c_i \qquad (32)$$

The error we use for modulation is the TD-error which is locally accumulated for every node in the DCS network. The lateral connection structure of the DCS is adapted with a learning rule derived from (12), again refer to [7] for details. A new neural unit (rule) is inserted whenever the distance to the current best matching unit is too large. At most N_{max} units are inserted.

4.2   Experiments

In order to test the applicability of our learning controller to collision avoidance with the TRC Labmate the simulated Labmate was placed in the training environment depicted in Figure 2. The Labmate was then allowed to drive around until either the distance to an obstacle dropped below $20cm$ or 200 time steps elapsed, ending a trial. In the former case, an orientation behavior is triggered which causes the Labmate to rotate until the front sensors indicate free space. In the latter case the Labmate is stopped and rotated for a random angle (to prevent it from staying on a closed trajectory all the time). Since we want to test the performance of the controller independent of incorporated prior knowledge the controller started with only one fuzzy rule:

if $(s_0 \approx 5000) \wedge \cdots \wedge (s_7 \approx 5000)$ then
  $(\mu_v(s) \approx 400) \wedge (\sigma_v(s) \approx \sigma_v^0) \wedge (\mu_\omega(s) \approx 0) \wedge (\sigma_\omega(s) \approx \sigma_\omega^0) \wedge (V(s) \approx 0)$,
stating that if all sensor readings are about $5m$ the Labmate should drive forward (zero angular velocity μ_ω) with velocity $\mu_v = 400 \; cm \, s^{-1}$. The certainty values for forward velocity and angular velocity $(\sigma_v, \sigma_\omega)$ were set to small initial values $(\sigma_v^0, \sigma_\omega^0)$. As immediate reinforcement we used the difference between evaluations of two succeeding situations, $r_t = \Phi(s_{t+1}) - \Phi(s_t)$, with $\Phi : \mathbb{R}^8 \to \mathbb{R}$ the evaluation function of a sensory situation. In addition the Labmate was given a high negative reinforcement signal if it had approached an obstacle within less than $20cm$.

For a typical run, Figure 13 shows the Labmate at the beginning of training in the training environment. Figure 14 shows collision free navigation of the Labmate in a test environment after training phase. End of training is indicated by the averaged TD-error approaching a minimum, the averaged reinforcement approaching its maximum and - of course - avoidance of collisions. Plots for the TD-error and the reinforcement (both averaged over 100 trials) are depicted in Figure 15. In our experiments training took between 1000 and 10000 trials, taking (on average) a longer time when sonar sensors with a characteristic beam

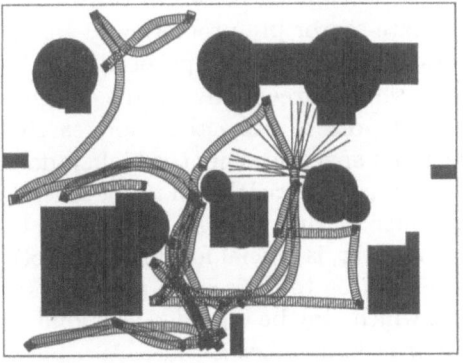 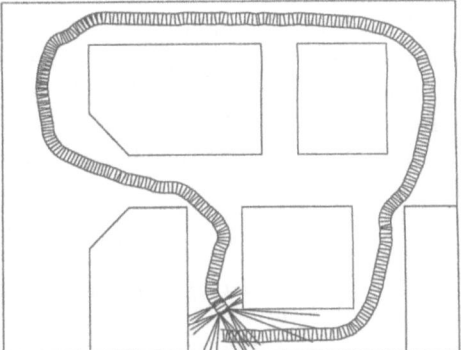

Figure 13: start of training, training en-   Figure 14: end of training, test environ-
vironment   ment

width of 20° and 5% noise were simulated than simulating idealized sensors (0°
beam width) without noise. However, the difference between these two types
of simulated sensors turned out to be surprisingly small. At most $N_{max} = 100$
neural units (rules) have been utilized. No effort has been spent on parameter
optimization.

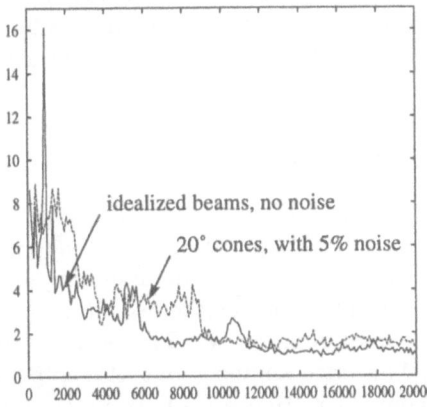 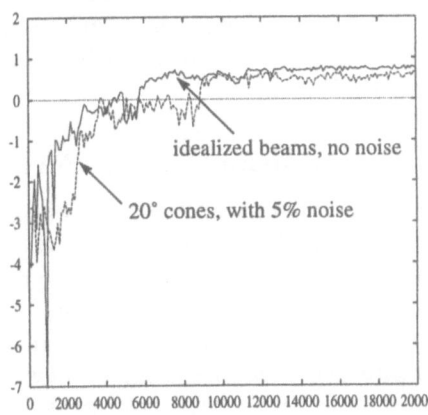

Figure 15: TD-error (right) and reinforcement (left) versus number of trials

5 Sensor Preprocessing

In most of the work on fuzzy logic control of mobile robots, sensor pre-processing
is either overlooked or highly simplified, assuming that the sensors deliver exact
(true) values [14, 15, 16, 17, 18]. On the contrary, we argue that it deserves
equal attention as the controller whose performance crucially depends on the
accuracy of the sensors.

There are two main reasons that necessiate sensor pre-processing in sonar-based fuzzy logic control. First, due to crosstalk, bad reflection properties of the environment and shielding problems, sonar sensors frequently produce highly erroneous readings which must be filtered out prior to using them for reactive control. Second, when the dimension of the input space becomes too high, it does no longer make sense to treat each input as a separate variable in fuzzy control. Besides the exponential growth of the number of rules it will be difficult to attribute a meaning to every input (which, however, is crucial for fuzzy control). In case of more than eight sonar sensors the solution to the second problem is a meaningful grouping of the sensor readings, which may be regarded as a kind of abstraction from the sensor readings using prior knowledge.

Such abstraction is not unique to the fuzzy control but has been addressed in connectionist inductive learning too. For instance, multi-layer networks have been trained in mobile robot navigation tasks, where the hidden layers construct a generalized intermediate representation of the input by supervised learning, [19, 20]. Likewise, reinforcement learning techniques have been employed to generalize the input space by recursively partitioning the state space based on the individual bit relevance [21]. However, a common characteristic of the above generalization techniques is that input uncertainties are not considered and domain knowledge is largly ignored.

The work in this section proposes a sensor pre-processing method for our TRC mobile robot. It combines *domain knowledge* and Kalman filtering to condense the sensory data and to cope with the uncertainties of the readings. The proposed pre-processing has been successfully tested on the real robot using an FRBF controller.

5.1 Partitioning the Perceptual Space

Instead of eight sensors in the previous sections (see Figure 4) in this section we utilize 10 sonar sensors. The sensors cover a total frontal angle of 120 *degrees* and are pre-programmed to measure a distance up to $2m$. Even assuming ideal sensors and the simplest output (binary output), the number of fuzzy rules to cover all input conditions is in the order of thousands! This is not only prohibitive from the point of view of the kowledge engineer but also imposes limitations on the reactivity of the fuzzy controller. Hence we have partitioned the ten ultrasonic sensors into five regions (Figure 16) corresponding to the physical geometry of the agent. These are: `right corner`, `left corner`, `right`, `front`, and `left`. In order to account for the beam angle of the sonars and the fuzzy nature of the five regions we adopt sensor *overlapping* across neighboring regions. The sensor arrangement and the overlapping perceptual regions are shown in Figure 16.

Similar to us, Reignier [17] also partitions the sensors into regions. Yet in order to determine the depth value of each region he only relies on a single sensor (the sensor that has minimum depth) . However, from our experience with the sonars of the Labmate the quantitative value of an individual sensor is not reliable at all. Instead, all sensors in a region must be taken into consideration

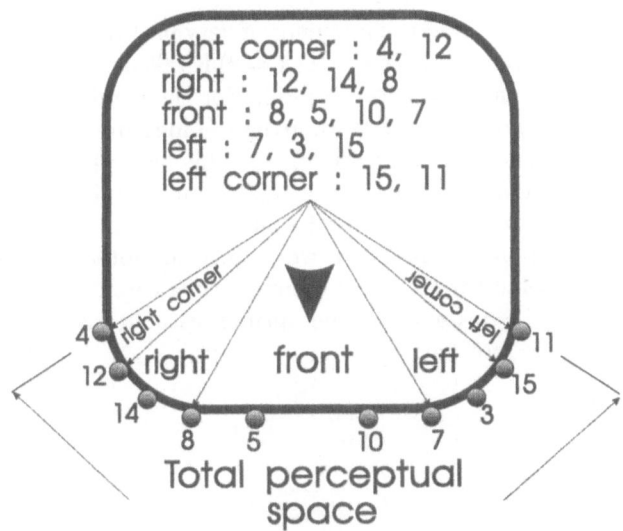

Figure 16: Overlapping perceptual regions

and filtering over time is advisable as well. In [20] we therefore partitioned the sensors into regions and refired the sonars multiple times at each perceptual cycle to gather data from which we estimated the depth of each region. However, multiple firing introduces noise in the system through sensor *crosstalk* and is time consuming.

Rather than relying on the quantitaive value of an individual sensor or firing the sonars repeatedly, we nowadays employ a cascade of two filters (Figure 17) and a sliding window of size 3 to hold the present and the past two measurement profiles. Following partitioning, the sensor values of each region are passed through a median filter (spatial filtering), which gives as output a single measurement of the depth of a region. The median filter estimates the current *measured depth* $\mathcal{Z}_{j,t}$ of a region j covered by N_j sensors using

$$\mathcal{Z}_{j,t} = median(\mathcal{S}_{1,t}^j, \mathcal{S}_{2,t}^j, \ldots, \mathcal{S}_{N_j,t}^j) \ . \tag{33}$$

where $\mathcal{S}_{i,t}^j$, $i = 1 \ldots N_j$, is the reading at time t of sensor i located in region j, N_j the number of sensors in region j and $\mathcal{Z}_{j,t}$, $j = 1 \ldots 5$ the measured depth of region j at time t. However, the measured depth $\mathcal{Z}_{j,t}$ is still noisy and too unreliable for reactive control[5]. Therefore, a Kalman filter is employed to further process the measured depth.

5.2   Kalman Filter Formulation

The proposed Kalman filter operates on the present and past measurement profiles $(\mathcal{Z}_{j,t}, \ldots, \mathcal{Z}_{j,t-n})$, stacked in the sliding window, to estimate the current true

---

[5]When these values are used to generate control commands the robot is seen moving arbitrary.

depth $\mathcal{D}_{j,t}$ of a region j. To avoid the influence of very past measurements on the present estimate only a limited window size ($n = 2$) is taken. Because a *Bayesian* viewpoint is adopted, we need to select a model for the conditional probability density function (CPDF) of the true depth given the measured depth $\mathcal{P}(\mathcal{D}_j/\mathcal{Z}_j)$ that best fits the data generated by real world. In this paper a Gaussian[6] CPDF is chosen. The main motivations for making this assumption is that the Kalman filter so designed is optimal with respect to virtually any criterion that makes sense [22]. As our viewpoint is Bayesian, we require the filter to propagate the assumed CPDF from some time $t - n$, for some arbitrary n, up to the present time t. Once the CPDF is propagated the optimal estimate is computed using the *maximum likelihood* criterion.

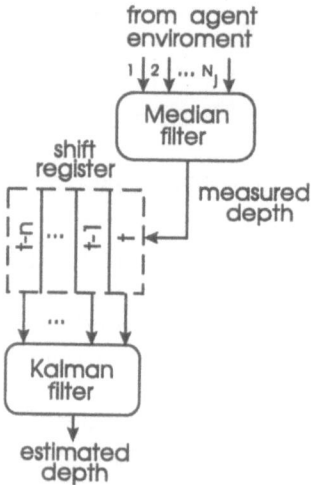

Figure 17: The proposed sensor pre-processor

The Kalman filter algorithm is tailored to suit the agent at hand. To proceed with the algorithm, at each perceptual time t the filters in each region $j = 1 \ldots 5$ are initialized by estimating the parameters of the Gaussian CPDF, mean (μ_j) and variance (σ_j^2). We estimate the mean by equating it with the measured value at time $t - n$ i.e,

$$\mu_{j,0} = \mathcal{Z}_{j,t-n}; \quad j = 1 \ldots 5 . \tag{34}$$

and the variance $\sigma_{j,0}^2$ with the measurement variance $\tilde{\sigma}^2$ of the sonars.

To compute the measurement variance, we have picked a sensor at random[7] and placed the sensor in different environments and at different orientations and depths that can be faced by the robot when it is in operation (such as corners, corridors, doors edges, walls, free ways, ...). For all environments and depths, the

---

[6]There is no mathematical or experimental prove that guarantees a Gaussian noise distribution in ultrasonic sensors.

[7]All the sensors are of the same Polaroid type.

sensor was fired and the true (d_i) and measured (r_i) depths were recorded. After recording $N = 1000$ (d_i, r_i) pairs, the measurement variance $\tilde{\sigma}^2$ is computed as

$$\tilde{\sigma} = \sqrt{\frac{1}{N} \sum_{i=1}^{N} (d_i - r_i)^2} = 137mm \; . \tag{35}$$

Yet this value turned out to be too low to represent the actual measurement variance of the sensors when fired one after the other on the moving robot. Hence we multiplied it by a factor of 2.5 to obtain $\tilde{\sigma} \approx 350$. At the beginning of the updating algorithm the statistical variance is set to this measurement variance, i.e,

$$\sigma_{j,0}^2 = \tilde{\sigma}^2 \quad j = 1 \ldots 5 \; . \tag{36}$$

With (34) and (35) the CPDF of each region is defined. The next step is to propagate this CPDF forward up to time t. Inherently our system is dynamic, i.e. agent position and hence sensor values change with time. Therefore, the dynamic Kalman filter best suits our scenario, yet this filter requires a model for the rate of change of the sonar return. For a situated agent, this change depends among other things on: the speed and rotation of the robot, the direction of motion, the environment and its acoustic properties, the dynamic properties of each sensor, the position of the sensors on the robot and the frequency of sensor crosstalk. Looking at the parameters involved, it is extremely difficult to come up with a clean mathematical model of the form of (37) and (38) for the dynamic filter:

$$\dot{\mathcal{X}}(t) = \mathcal{A}(t)\mathcal{X}(t) + \mathcal{B}(t)\mathcal{U}(t) + \mathcal{V}(t) \; . \tag{37}$$

$$\mathcal{Z}(t) = \mathcal{C}(t)\mathcal{X}(t) + \mathcal{W}(t) \; . \tag{38}$$

Here matrices $\mathcal{A}(t)$, $\mathcal{B}(t)$ and $\mathcal{C}(t)$ are system time varying coefficients incorporating all the above mentioned parameters, vectors $\mathcal{X}(t)$ and $\mathcal{Z}(t)$ are estimated and measured depths respectively, and $\mathcal{V}(t)$ and $\mathcal{W}(t)$ are system and measurement noises respectively.

Because of lack of the above system coefficients, a linear recursive Kalman filter is employed, and the CPDF is updated only at discrete time steps, when a measurement value is available. At each update step $i = 1, \ldots, n$ and for any perceptual region $j = 1 \ldots 5$, the updating algorithm is given by :

- compute Kalman gain:

$$\mathcal{K}_{j,i} = \frac{\sigma_{j,i-1}^2}{\sigma_{j,i-1}^2 + \tilde{\sigma}^2} \; . \tag{39}$$

- update mean:

$$\mu_{j,i} = \mu_{j,i-1} + \mathcal{K}_{j,i}(\mu_{j,i-1} - \mathcal{Z}_{j,t-n+i}) \; . \tag{40}$$

• update variance:

$$\sigma^2_{j,i} = (1 - \mathcal{K}_{j,i})\sigma^2_{j,i-1} \ . \tag{41}$$

Figure 18 shows how the parameters of the CPDF, μ_j and σ^2_j, vary at each update. At the last update, we have the CPDF of the estimated depth given the present and the past two measured values, $\mathcal{P}(\mathcal{D}_{j,t}/\mathcal{Z}_{j,t-2}, \mathcal{Z}_{j,t-1}, \mathcal{Z}_{j,t})$. Once this CPDF is determined, the *maximum likelihood* criterion is used to extract the best estimate from the CPDF, i.e,

$$\begin{aligned} \mathcal{D}_{j,t} &= \ max \ \mathcal{P}(\mathcal{D}_{j,t}/(\mathcal{Z}_{j,t-2}, \mathcal{Z}_{j,t-1}, \mathcal{Z}_{j,t})) \\ &= \ \mu_{j,2}; \quad j = 1 \ldots 5 \ . \end{aligned} \tag{42}$$

We have implemented a separate Kalman filter according to Figure 17, (34), (35) and (39–42) for each region. Taken together they define our *sensor preprocessing* stage.

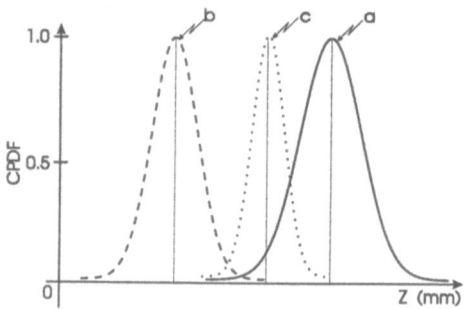

Figure 18: Variation of CPDF of $\mathcal{D}_j$ based on (a) $\mathcal{Z}_{j,t-2}$, (b) $\mathcal{Z}_{j,t-2}, \mathcal{Z}_{j,t-1}$, (c) $\mathcal{Z}_{j,t-2}, \mathcal{Z}_{j,t-1}, \mathcal{Z}_{j,t}$

5.3   Experimental Results

In order to test the preprocessing stage, the robot was placed at a distance of $2m$ in front of a wall. After firing all the sonars the robot is set to move against the wall at a constant velocity. While it was moving, we kept on recording the readings of the sonars in the front regions (sensors $5, 7, 8$ and 10) until the robot approached the wall. Afterwards we applied our as well as Reignier's pre-processing algorithm to the data gathered. Figure 19 shows a plot of both results over time. Clearly, our pre-processing (broken line) provides the better estimate. In particular, notice how the Kalman filter holds (sustains) the depth estimate at a relatively high value with only little oscillations when the robot is far from the wall.

Apart from this off-line test, the proposed sensor pre-processing has been used with an FRBF controller on the actual TRC robot. It enabled the robot to move in our office environment and to pass even narrow doors.

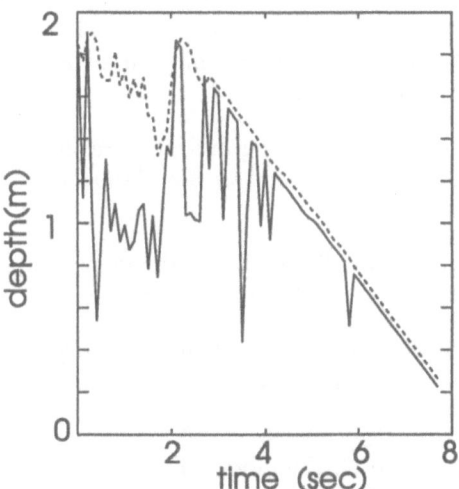

Figure 19: Performance test of the proposed sensor preprocessor

References

[1] J. Bruske and G. Sommer, "Dynamic cell structure learns perfectly topology preserving map," *Neural Computation*, vol. 7, no. 4, pp. 845–865, 1995.

[2] D. Nauck, F. Klawonn, and R. Kruse, *Neuronale Netze und Fuzzy-Systeme*. Braunschweig/Wiesbaden: Vieweg, 1994.

[3] B. Kosko, *Neural Networks And Fuzzy Systems*. Englewood Cliffs, New Jersey: Prentice Hall, 1992.

[4] J. Jang and C. Sun, "Functional equivalence between radial basis function networks and fuzzy inference systems," *IEEE Transactions on Neural Networks*, vol. 4, no. 1, pp. 156–158, 1993.

[5] I. Ahrns, "Ultraschallbasierte navigation und adaptive hindernisvermeidung eines autonomen mobilen roboters," Master's thesis, Inst. f. Inf. u. Prakt. Math., CAU zu Kiel, 1996.

[6] T. Martinetz and K. Schulten, "Topology representing networks," in *Neural Networks*, vol. 7, pp. 505–522, 1994.

[7] I. Ahrns, J. Bruske, and G. Sommer, "On-line learning with dynamic cell stuctures," in *ICANN'95*, pp. 141–146, 1995.

[8] M. J. Mataric, "Integration of representation into goal-driven behavior-based robots," *IEEE Transactions on Robotics and Automation*, vol. 8, no. 3, pp. 304–312, 1992.

[9] J. Buhmann, W. Burgard, *et al.*, "The mobile robot rhino," *AI Magazine*, vol. 16, no. 1, 1995.

[10] J. Bruske, I. Ahrns, and G. Sommer, "An integrated architecture for learning of reactive behaviors based on dynamic cell structures," *Robotics and Autonomous Systems*, 1997. in press.

[11] R. S. Sutton, "Reinforcement learning architectures for animats," in *Proc. of the First International Conference on Simulation of Adaptive Behavior*, pp. 288–296, MIT Press, 1990.

[12] R. S. Sutton, "Learning to predict by the methods of temporal differences," *Machine Learning*, vol. 3, no. 1, pp. 9–44, 1988.

[13] R. J. Williams, "Simple statistical gradient-following algorithms for connectionist reinforcement learning," *Machine Learning*, vol. 8, pp. 229–256, 1992.

[14] T. Takeuchi, Y. Nagai, and N. Enomoto, "Fuzzy control of a mobile robot for obstacle avoidance," *Information Science*, vol. 45, pp. 231–248, 1988.

[15] K. T. Song and J. C. Tai, "Fuzzy navigation of a mobile robot," in *Proceeding of the 1992 IEEE/RSJ International Conference on Intelligent Robots and Systems*, (Raleigh), pp. 621–627, 1992.

[16] C. C. Jou and N. C. Wang, "Training a fuzzy controller to back up an autonomous vehicle," in *IEEE International Conference on Neural networks*, (San Francisco), pp. 923–928, 1993.

[17] P. Reignier, "Fuzzy logic techniques for mobile robot obstacle avoidance," *Robotics and Autonomous Systems*, vol. 12, pp. 143–153, 1994.

[18] A. Saffiotti, H. Ruspini, and K. Konolige, *Using Fuzzy Logic for Mobile Robot Control*. Kluwer Academic, 1997. Forthcoming.

[19] D. A. Pomerleau, *Neural Network Perception for Mobile Robot Guidance*. Kluwer, 1993. Norwell, MA.

[20] G. Hailu, "Distributed fuzzy and neural network based navigational behaviours," Technical Lab. Report H/1996, CAU, Cognitive Systems Laboratory, 1996.

[21] S. Mahadevan and J. Connell, "Automatic programming of behavior-based robots using reinforcement learning," *Artificial Intelligence*, vol. 55, pp. 311–365, 1992.

[22] P. S. Maybeck, *The Kalman Filter: An Introduction to Concepts*, pp. 194–204. Springer Verlag, 1994.

Chapter 9

TRENDS IN EVOLUTIONARY ROBOTICS

Lisa Meeden
Computer Science
Swarthmore College
Swarthmore, PA
USA

Deepak Kumar
Department of Mathematics
Bryn Mawr College
Bryn Mawr, PA
USA

A review is given on the use of evolutionary techniques for the automatic design of adaptive robots. The focus is on methods which use neural networks and have been tested on actual physical robots. The chapter also examines the role of simulation and the use of domain knowledge in the evolutionary process. It concludes with some predictions about future directions in robotics.

1 Introduction

To be truly useful, robots must be adaptive. They should have a collection of basic abilities that can be brought to bear in tackling a variety of tasks in a wide range of environments. These fundamental abilities might include navigation to a goal location, obstacle avoidance, object recognition, and object manipulation. However, to date, this desired level of adaptability has not been realized. Instead, robots have primarily been successful when deployed in constrained environments to perform deterministic tasks. The result has been that robots have had very limited, task-specific competencies which do not generalize to new situations.

The main advantage of imposing strict constraints within a robotic domain is to enable the use of a predictive model. This implies that a particular set of sensor readings can be accurately translated into a representation of the current state of the world. Furthermore, the result of doing a particular action can be known in advance of actually executing it. This allows a robot to plan its behavior.

Even working within these simplified conditions, a third of the cost of an industrial robotic system is its programming [1]. Yet, the most exciting potential applications involve much more complex and dynamic environments than have typically been attempted so far (for example outdoor, subsea, and other planet environments). We should expect that the burden on the human designer of control software will only increase as we try to move towards these more advanced applications. One hope is that the uncertainty and variability that occur in physical sensors and actuators might lessen as our hardware technology improves, thus alleviating the programming problems that arise. However, some experiments have shown that using higher resolution sensors introduces more variation, not less as one might expect [2].

The use of evolutionary computation is one of the most promising avenues for overcoming the bottleneck of the human engineer in the robot design process. In fact, it has been proposed that an evolutionary approach to the design of robots will eventually supercede design by hand [3]. The fundamental idea of this approach is to maintain a population of possible robot control architectures. The initial population is typically a collection of randomly configured architectures. Each architecture is evaluated according to an objective fitness measure and the better the robot performs using that architecture the more offspring it is allowed to produce in the next generation of the population. Over a number of generations, the fitness of the population increases and successful architectures are created. A human engineer must develop the evolutionary framework, but the actual design of the robotic systems is then automatically generated.

The most accurate fitness measure for a potential architecture is to allow it to control the actual robot operating in the physical world. Yet this can be a painstakingly slow process. Consider that each action can require a second of time to execute and that each fitness measure typically involves hundreds of actions. With a sizable population, the processing of a single generation may require an hour's time. When hundreds of generations are necessary to achieve reasonable results, one run of the evolutionary process may require several days. Two techniques have been employed to speed up the evolutionary process: the use of simulations for faster fitness evaluations and the seeding of the initial population with domain knowledge to decrease the overall number of generations required.

A central question when adopting the evolutionary computation approach is: What type of robot control architecture should be evolved? There are a number of options: high-level code [4], machine code [5], parameter settings for a hand-designed system [6], situation-action rules [7], and entire rule-based strategies [8]. Perhaps the most innovative direction, however, is the combination of evolutionary computation with artificial neural networks. Neural networks allow the evolutionary process to operate at a very low level, placing minimal constraints on the possible solutions. When a higher-level architecture is used designer bias is more prevalent [3]. In addition, neural networks are robust, noise tolerant and can be used for local learning starting from the points discovered by the global evolutionary process [9].

The remainder of this chapter is organized as follows. In Section 2, further background is given on evolutionary computation and neural networks. In Section 3, some specific research projects combining evolutionary computation and neural networks to control robots are examined. In Section 4, the current debate on the role of simulations is reviewed. In Section 5, the ways in which domain knowledge has been incorporated in evolutionary robotic systems is discussed. Finally in Section 6, speculations about future trends in robotics are explored.

2   Background

2.1 Evolutionary Computation

The majority of the current implementations of evolutionary computation descend from three branches of study: genetic algorithms, evolutionary programming, and evolution strategies [10]. Of these, the genetic algorithm approach developed by Holland, is the basis for most evolutionary robotics applications [11], [12]. Other related techniques that arose out of genetic algorithms are classifier systems [13] and genetic programming [14].

All evolutionary computation methods attempt to mimic the process of natural evolution. The general structure of an evolutionary computation algorithm is shown in Figure 1 and depends on three operators: selection, recombination, and mutation. *Selection* is usually implemented as a probabilistic process using the relative fitness of an individual to determine its selection probability. In this way, fitter individuals are more likely to participate in producing the next generation. *Recombination* is the means of mixing the genetic material of two parents to produce an offspring. *Mutation* creates random alterations in the genetic material of an offspring.

```
t=0
randomly initialize population(t)
determine fitness of population(t)
repeat
  select parents from population(t)
  recombine and mutate parents creating population(t+1)
  determine fitness of population(t+1)
  t = t+1
until best individual is good enough
```

Figure 1. General Evolutionary Computation Algorithm

Consider the effects of each of these operators. Selection alone tends to fill the population with copies of the best individual from the initial population. Mutation alone induces a random walk through the search space. Selection and mutation (without recombination) creates a parallel, noise-tolerant, hill-climbing search. Selection and recombination (without mutation) tends to cause the process to converge on good, but sub-optimal solutions. Together the three operators form a flexible, robust, global search method.

2.2 Artificial Neural Networks

One of the most common artificial neural architectures is a layered feed-forward network. Each unit in one layer is linked by weighted connections to each unit in the next layer. Various amount of activation are applied to the units in the initial layer, representing the input to the network. This activation then flows across the connections to higher layers, with the weights mediating the amount of activation that is passed on to successive units. The final pattern of activation present on the last layer is considered to be the output produced.

A supervised learning algorithm such as back-propagation can be repeatedly applied to adjust the weights of a network enabling it to learn to associate arbitrary pairs of input and output patterns [15]. If the input pattern is interpreted as representing sensory information and the output pattern as an action, a network can be used to control a robot. As a result of this training process, the network learns to re-code the incoming sensory patterns into new patterns at the intermediate layers so that the appropriate output action is produced. These intermediate layers are termed *hidden* because they do not have direct access to the environment.

One drawback of the standard feed-forward architecture is that it is difficult to represent time in an efficient manner. Certainly for controlling a robot timing information is crucial. A common method of accommodating time has been to represent time as space. This entails extending the input layer to act as a buffer of the past. A better approach is to deal with time implicitly rather than explicitly by using a recurrent architecture. One such recurrent architecture, called an Elman network, allows every unit in a hidden layer to have weighted connections to every other unit in the layer, including itself [16]. This gives the network a memory of its own internal re-codings of its past sensory inputs. A more radical recurrent architecture abandons a layered topology and allows units to connect to one another in arbitrary ways.

In Section 3, we will see examples of feed-forward, Elman, and arbitrary network architectures being employed to control robots.

2.3 Combining Evolutionary Computation and Artificial Neural Networks

In applying evolutionary computation to neural networks for the purpose of robot control, two main methods have been used. The first method is to fix the topology of a network architecture and to then use evolutionary computation to determine the weights. Unlike back-propagation, which is a gradient descent procedure for finding an appropriate set of weights, evolutionary computation should avoid getting stuck in local minima. The second method is to allow evolutionary computation to actually determine aspects of the network's topology. The weights may then be set through a separate learning algorithm or can also be determined by evolutionary computation.

In Section 3, we will see examples of both of these approaches, evolved weights and evolved topologies, being used to develop robot controllers.

3   Examples of Evolved Neural Network Robot Controllers

Conducting evolution on physical robots is a time consuming process, and as a result most of the applications attempted so far have been fairly modest in scope. Often, for reasons of practicality, the robots used are quite small. This allows the task environment to be set up on a desktop with the robot tethered to a computer for data collection and tethered to an electrical outlet for power. One popular platform for conducting evolutionary experiments is the Khepera robot [17]. Khepera is circular in shape and miniature (diameter 55 mm, height 30 mm, and weight 70g). It has two DC motors which power two wheels. It's standard sensory apparatus consists of eight infra-red proximity sensors.

Of the five applications described below, the Khepera was used in two: battery recharging and trash collection. Another small robot, called the Gantry, was used for locating visual targets. A miniature car, called carbot, was used for light seeking and avoiding. Only the driving task involves a full size robot, a car called NAVLAB, which is part of an ongoing project at Carneigie Mellon University [18], [19].

3.1 Battery Recharging [20]

The environment was an empty rectangular arena with a gray floor. A light emitter was placed over one corner, and under this light a circular patch of the floor was painted black. This represented the battery recharging area. When the robot happened to pass over this black area its virtual battery would be instantly recharged. The Khepera robot was equipped with some additional sensors for this task. First it had three light sensors, two of which gathered ambient light, and one which pointed down at the floor. Second it had a virtual battery level sensor.

The goal of the evolutionary process was to determine an appropriate set of weights for a fixed Elman-style recurrent network with twelve input units for the sensors, five hidden units, and two output units for the motors. The process was run continuously for ten days with each generation lasting approximately three quarters of an hour.

Each individual set of weights was evaluated by a very simple fitness function that consisted of two components: one to maximize speed and the other to avoid the walls.

When the robot was in the recharging area, fitness was defined to be 0. Notice that there is no explicit reward for recharging the battery; in fact in terms of direct fitness it is detrimental to spend time in the recharging area. However, there is an implicit benefit to recharging. A fully charged battery allowed the robot to move for 50 time steps, and each individual was allowed an upper limit of 150 time steps for evaluation. Therefore by recharging the battery, the robot could gain more time to accumulate fitness.

Analysis of the resulting behavior revealed that the robot exhibited very different strategies depending on its battery level. When the battery level dropped to about one third full charge, the robot began executing a trajectory that led it to the recharging area. Otherwise, it moved at nearly full speed along a slightly bended trajectory that avoided the recharging area.

Although a very general fitness function was used, the evolved robot controller was able to locate the battery recharging area and to instigate timely homing maneuvers when the battery level was low. To extend this generality even further, Floreano and Mondada suggest that it would be interesting to redo this experiment but eliminate the fitness function entirely and simply select those individuals that live longer. This would make the artificial evolution process more similar to natural evolution.

3.2 Trash Collection [21]

In order to pickup trash, the Khepera robot was equipped with a gripper module. The environment was a rectangular arena containing five pieces of randomly distributed trash (white cardboard cylinders). The robot's task was to find these targets, pick them up, carry them to the boundary of the arena, and drop them outside.

Nolfi compared five different architectures, the best of which was able to determine how to modularize the task. It was a two-layer feed-forward network where the value of each motor output depended on a competition between two separate modules. The number of available modules was fixed at two per output node, but the interactions between the modules was determined by the adapted weights.

The fitness measure primarily depended on how many targets were successfully released outside the arena. To help bootstrap the process, the fitness measure also included a component to reward a robot's ability to simply pickup targets. It proved difficult for the robot to learn how to react to new targets when it was already carrying a target. To alleviate this problem, the training experience was manipulated to make this type of occurrence more frequent.

During the training process, the controllers were evaluated in a simulator. In spite of this, each evolution run still required approximately ten hours. After training was complete, the resulting control systems were downloaded into the physical robot and tested in the real environment.

To succeed at this task, the ability to distinguish between walls and targets is of crucial importance. Some errors made by the robots revealed that this was not a simple distinction, such as attempting to grasp walls or releasing a target over another target. The emergent modular architecture created modules specifically tuned to making this distinction under confusing sensory situations. The capacity to produce sharp switches in strategy in response to fine-grained sensory differences is a key to complex behavior.

3.3 Locating Visual Targets [22]

Several visual tasks were tried, including locating a large static target, locating a smaller static target, tracking a moving target, and locating a triangle target in the presence of a competing rectangle target. For these experiments a specialized piece of robotic equipment was developed called the Gantry robot. Instead of wheels, the robot is suspended from a gantry frame that allows translational movement in the X and Y directions. The robot is 150mm in diameter and can rotate. It is equipped with a camera pointed down at a mirror which is inclined at 45 degrees as well as several bumper sensors.

In this case, the evolutionary algorithm searched for both a network architecture and a visual morphology. Each network had a fixed number of input nodes (for sensors) and output nodes (for motors), but the number of hidden nodes and the number and type of connections (either excitatory or inhibitory) was variable. Connections were allowed to be recurrent between any layer. Furthermore, rather than feeding a raw camera image to the controller, they allowed the method of sampling this image to be evolved along with the network. They achieved this by designating a set of possible receptive fields within the image.

The visual tasks involved finding a light colored target within the dark colored environment. The fitness measure was a function of the distance of the robot from the desired target--the closer the robot to the target the higher its fitness. For each possible architecture and visual morphology, the robot was given four trials starting from the same position but using different orientations. The final fitness was the worst score from these four trials. This allowed them to terminate trials as soon as they bettered a previous trial and thus improved the evaluation time.

Starting the evolutionary process from initially random populations proved to be too slow. Instead they selected one of the members of this initial random population that displayed "interesting" behavior and then cloned it to create a new population. Working with a converged population rather than a random population has been termed Species Adaptation Genetic Algorithm or SAGA [23]. From this starting point, the evolution run was able to develop good solutions within 10 to 15 generations with each generation taking about one and a half hours.

First the robot was evolved to successfully locate large (150cm wide) static targets. The final population from this experiment was used as the starting point for the next experiment of locating smaller (22cm wide) static targets. In less than 10 generations the population adapted successfully to the harder task. Again his new population was recycled and used as the starting point for the even harder task of tracking a moving target. Working in this incremental fashion it should be possible to gradually build up quite complex behaviors (see also [24]). Also allowing the morphology of the robot to be developed along with the controller may simplify the task solution considerably.

3.4 Seeking and Avoiding Light [25]

Carbot is a modified toy car which is approximately 15cm wide and 23cm long and is equipped with two light sensors and several bumper sensors. It was placed in a rectangular arena with a light in one corner. Its task was to constantly keep moving, avoid the walls, and respond appropriately to a light goal. When the goal was positive, carbot had to seek out the light until a maximum light reading was obtained. Once this was accomplished the goal automatically switched to a negative value, indicating that carbot had to avoid the light until a minimum light reading was obtained. The goal varied in this periodic manner throughout the task, seeking was immediately followed by avoiding and so on.

A local and a global method of reinforcement learning were compared for training carbot at this task--a special form of back-propagation and an evolutionary algorithm. The topology of the network was fixed to be an Elman network with seven inputs for the sensors and goal, five hidden units, and four output units for the motor settings. The aim of both methods was to find an appropriate set of weights. The adaptation process was conducted on a simulation and the best architectures were tested on the actual robot.

Statistical analyses of the results revealed several quantitative differences between the two learning methods. The back-propagation algorithm out-performed the evolutionary algorithm in the original task. However, the evolutionary algorithm out-performed the back-propagation algorithm when the task difficulty was increased either by removing the explicit goal (but keeping the periodic structure) or by removing immediate reinforcement feedback. Perhaps even more interesting were the qualitative differences in the behaviors produced by the two methods. Being a local method, back-propagation was more sensitive to the moment-to-moment changes in the environment and thus used the explicit goals to develop unique strategies tuned to each goal. In fact when no explicit goal was present, back-propagation trained networks sometimes created their own goal-like units in the hidden layer. In comparison, the evolutionary algorithm tended to develop a single overall strategy that was applicable to both goals. More importantly, the evolutionary algorithm's ability to find good strategies was quite robust across the experimental variations.

The respective strengths and weaknesses of these two methods are clearly complementary, suggesting that some hybrid of the two could be the most effective method. Because the evolutionary algorithm globally samples the entire space of alternative solutions while back-propagation locally searches the immediate neighborhood of a solution, the most straight-forward form of hybrid would be to allow the evolutionary algorithm to find a good starting point in the weight space and then use back-propagation to do the fine tuning. As in nature, the global evolutionary method can determine a good gross solution which the local learning method can then adjust to the current environmental conditions.

3.5 Driving [26]

NAVLAB is an autonomous land vehicle that has been operated in a wide variety of domains including dirt roads, bike paths, two-lane suburban neighborhood streets, and divided highways. A NAVLAB controller must determine an appropriate steering angle when given a video image from a camera mounted on the front of the car. Unlike the previous tasks, for this task there is a clear "right" answer for every situation as determined by human drivers.

A number of different network architectures have been explored for solving this driving task, the most successful of which is a three-layer feed-forward topology. It has a 2-D input retina for the video images, a small hidden layer, and a gaussian representation of the steering angle across thirty output units ranging from "sharp left" to "straight ahead" to "sharp right". Previously back-propagation had been used to determine the weights for these network controllers. Baluja set out to discover whether evolutionary computation could develop better controllers for this task.

Suppose that a *maximal* network describes the maximum connectivity of the networks to be evolved. Then through evolution, different topologies can be tested by selecting which of these possible connections to enable. If a connection is present, then a weight for it is determined as well. The maximal network for these experiments contained a 15x16 input retina fully connected to a five unit hidden layer fully connected to a thirty unit output layer, all strictly feed-forward.

In order to reduce search times, Baluja created a novel evolutionary method called Population-Based Incremental Learning (PBIL). This algorithm requires that individuals in the population be represented with a binary alphabet (network weights can be represented as binary numbers if restricted to a range of values). The goal of PBIL is to produce a real-valued probability vector which can be considered a prototype for highly fit individuals in the population. By sampling this single probability vector an entire population can be stochastically produced and then tested. Based on the test results the values are adjusted to push the probability vector towards the best individual and away from the worst individual. In addition, a mutation operator is used to update the probability vector directly, by shifting each value a small positive or negative amount. Unlike a genetic algorithm, PBIL does not use any recombination operator.

To test possible networks for the driving task, a training set of 1000 images and correct steering angles were collected and saved. From this set, 100 examples are randomly selected for each network evaluation. The network which obtained the lowest error was designated best for that generation and the network with the highest error the worst.

The final networks evolved using the PBIL method kept only about half of the possible connections allowed in the maximal network. When compared with maximal networks trained with back-propagation, the PBIL networks showed a 13% reduction in error. Thus PBIL produced more space efficient and more accurate controllers. However this gain has a cost; to complete an entire PBIL run requires about an hour while a back-propagation run requires only a few minutes of processing time.

3.6 Summary of Examples

With the battery recharging task, Floreano and Mondada demonstrated that interesting, complex behavior can be obtained without an overly explicit fitness function. They further suggested that rather than employing a human engineered measure of some kind, ultimately the best measure may be the most general one-- survival of the fittest.

In contrast to this call for more implicit fitness measures, Harvey, Husbands, and Cliff argued that complex behavior may best be obtained through a set of well designed incrementally harder fitness tests. Using related visual location tasks, they showed that by beginning each evolution run from a converged population rather than a random one, a faster more focused exploration was produced.

Another important aspect of Harvey, Husbands, and Cliff's experiments was that the visual morphology of the robot was evolved along with the controller. Because the human sensory apparatus is quite different from a robot's sensory apparatus, the way in which a human might consider processing a visual image will probably not translate well into a robot. Instead, by allowing the evolutionary process to operate on the sensory apparatus, a more efficient robot-based solution can emerge.

A similar conclusion was drawn by Nolfi with respect to modularization. Through a comparison of a number of network architectures for a trash collection task, Nolfi found that providing a network with modularization options was extremely beneficial. Again because of the differences between humans and robots in sensory capabilities, the way in which a human might subdivide a task may not translate well into a robot. Allowing the evolutionary process to consider how to break up a task can lead to simpler robot-centered solutions.

In the light seeking and avoiding task, Meeden demonstrated that an evolutionary algorithm is a robust method for determining a good gross solution to a robot control problem. She suggested that such a solution can be improved and fine tuned through additional training with a local learning method such as back-propagation. A hybrid of this kind can be robust across large environmental changes and yet sensitive to subtle features.

Finally in Baluja's studies, a fast, new evolutionary computation method called PBIL was employed to create controllers for driving. Both the topology and the weights of the networks were determined by the evolutionary process. The results demonstrated that automatically designed controllers out-performed hand-designed controllers.

One question that arises from these studies is: How large a role should the human designer play in shaping the robot's behavior through the fitness function? Should our models of evolution try to be more true to natural evolution? Or should we as engineers try to influence the evolutionary process more directly?

A related question that also emerges from these studies is: What aspects of the robot control system should be manipulated by the evolutionary process--the parameters, the architecture, or the robot itself? It appears that the more features that are accessible to the evolutionary process then the more successful the adapted controllers will be.

4   Simulation

There is currently a hot debate among people trying to understand and reproduce intelligent agents, that could be stated as follows: Is the simulation a powerful enough tool to draw sound conclusions, or should a theory or an approach be tested on a real agent, i.e. a robot? [27]

Simulation in robotics, control theory, and AI has mostly been a complete waste of time. Of course there are certain cases in which simulation is inevitable ... what is at issue is whether results 'demonstrated' using simulation *only* should be accorded worth. We think not. [28]

However, it does appear that simulations are not quite the dead-end some had suggested. For simpler cases at least it has been shown that they can be made accurate enough. Their attractive qualities of speed and ease of data collection can then be made use of. [29]

It is frankly easier to use robots situated in the real world than it is to try to build some all encompassing super-simulation. [30]

What makes a robot distinct from any other artificial intelligence project is that one must actually deal with hardware and the intrinsic limitations that all physical sensing and acting systems have [31]. By using a simulation, one can sidestep these difficult hardware interactions completely, and this is what is at issue in the simulation debate. For example, the term "robot" is often used loosely to refer to a simulated "agent" or "animat" which may not have any physical counterpart upon which it is modeled [2]. In fact some such simulated "robots" could never be implemented in the real world because they depend on non-existent "sensors" to provide object-level information about the environment. The danger is that simulations simplify the learning problem too much by making the environment and the robot clean and predictable. Thus there is no guarantee that results obtained in simulation will transfer to the noisy real world. This is termed the *correspondence problem.*

Despite the difficulties simulations present they are still very attractive primarily because of their speed. An evolutionary experiment that takes several days on a physical robot may only require several hours on a simulated robot. This ability to obtain results quickly facilitates a more open-ended exploration of robot control ideas. In addition, simulated results are more easily collected, analyzed, and reproduced.

To lessen the correspondence problem, every effort should be made to keep simulations in close step with reality. Some specific suggestions have been made along this line: (1) base the simulation on carefully collected empirical data of a real physical robot; (2) add noise to the simulated sensory readings and the actuator outcomes; and (3) calibrate the simulation through tests on the real physical robot [32].

Calibration tests have provided interesting and somewhat contradictory results. In one case, neural network robot controllers adapted in simulation always performed better when tested in the real world than they had when tested in simulation [33]. The suspected cause of this improvement was that the physical robot's movements and sensor readings were not noisy in the same ways as the simulator's. The physical robot's experience was occasionally noisy while the simulator's experience was systematically noisy, and this was beneficial to learning. In another case, experiments were conducted to determine how the amount of simulated noise affects the correspondence between behavior evolved in simulation and then tested in simulation versus being tested in the real world [29]. Three noise levels were examined: no noise, observed noise, and double the observed noise. The results revealed that networks that evolved in an environment that is less noisy than the real world will behave more noisily in reality. Conversely, networks evolved in an environment that is noisier than the real world will behave less noisily in reality. Furthermore the correspondence between simulation and reality was maximized when the noise level of the simulation most closely matched reality.

In the first case additional noise appeared to be beneficial while in the second case it did not. Perhaps these contradictory results could be resolved with further experiments. It may be that twice the observed noise is too drastic a change to realize a benefit, whereas a smaller increment above observed noise would be helpful. This issue of the appropriate amount of noise is just one of the many open questions related to the use of simulations for adapting robots.

In doing evolutionary robotics, one has three options for evaluating possible control systems: use a physical robot, use a simulated robot, or use a hybrid of the two [9]. Using a physical robot is obviously the most desirable but may be too time consuming. Using a simulated robot leads to the correspondence problem. Using a hybrid approach, one can begin the evolutionary process on a simulated robot to quickly develop a high performing population. Assuming that the simulation was developed through close observation of the actual robot, this should provide a good starting point for the slower evolutionary process on the physical robot. Also a simulated robot can be used to quickly prune the set of possible experiments one may want to eventually conduct on a physical robot [34]. For these reasons, the hybrid approach to evolutionary robotics may currently offer the best compromise between speed and accuracy.

5 Domain Knowledge

The role of domain knowledge in evolutionary robotics can be viewed as a continuum. At one extreme, robots can have complete domain knowledge and require no ability to adapt, while at the other extreme robots can have no domain knowledge and be seen as tabula rasa systems. At a recent workshop on robot learning, most of the research presented there incorporated a substantial amount of domain knowledge [35]. Many of the presenters argued that to construct a successful robotic system, learning must be limited in use to portions of the system where the designer's knowledge is too sketchy to engineer a solution. Learning from scratch was seen as too inefficient for any problem of reasonable complexity.

A middle ground along this continuum is an approach that treats knowledge acquisition as a cooperative effort between the human engineer and the robot itself [36]. Grefenstette developed an evolutionary method known as SAMUEL, which stands for Strategy Acquisition Method Using Empirical Learning [37]. SAMUEL is an evolutionary process that develops entire behaviors which are defined as sets of rules. The rules are expressed in a high-level language to make it easy to incorporate existing knowledge and to make it easy to understand the results of learning. The initial population of behaviors are not random but consist of a variety of rule sets including human generated ones and automatically generated variants of these. By seeding the initial population, it is hoped that the search space will be usefully constrained thus leading to faster search times (refer back to Section 3.3 for another example of this). Another technique used to constrain the search space, is the creation of virtual sensors. For instance, rather than basing rules on sixteen raw sonar values, one could create four virtual sensors--left, forward, right, and backward--which combine the raw values in a meaningful way.

Although beginning with a significant amount of domain knowledge seems more practical there may be a disadvantage. By imposing our perspective on the learning problem we may actually make it harder. Nolfi argues that a designer views a robot task from an observer's point of view or a *distal* perspective, but a robot must solve the task in terms of its sensory-motor system or a *proximal* perspective [21]. There may be no simple mapping from the distal perspective to the proximal perspective. For instance, consider the trash collection task described in Section 3.2. One distal description of this task is the following: (1) explore the environment avoiding walls; (2) recognize trash and approach it for pickup; (3) pickup trash; (4) move towards a wall avoiding other trash; (5) release trash over the wall. Recall that the Khepera robot used in these experiments only had access to infra-red sensors. Using human vision (the distal perspective) it is a simple matter to distinguish a wall from trash, but using the robot's infra-red sensors (the proximal perspective) it is not. For the robot, these two objects can only be distinguished from a relatively small number of close positions. Thus a modular division of the task into distal subtasks such as "move towards a wall avoiding other trash" is not viable because trash and walls appear the same except from a very local view.

Artificial neural networks offer one of the most promising means for investigating robot control because they allow the task demands rather than the designer's biases to be the primary force in shaping the system's development. Yet the designer still has an important role to play in the evolutionary process which includes: determining what aspects of the neural network will be operated on by evolution, the input and output representations, the robot's physical characteristics (which could also be manipulated by evolution), and the robot's environment. After this adaptive process has been set in motion and a successful control system has been produced, its method must be dissected to understand the underlying control principles. The use of evolutionary computation with neural networks inverts the classical order of problem solving in which a high-level understanding comes first and closely guides the search for algorithms [38]. Through the evolution of neural network controllers a solution to the task emerges which is not simply a product of the designer's understanding of the domain.

6   Future Trends in Robotics

Hans Moravec likens today s robots to simple invertebrates in the global evolutionary sense [39]. He predicts that in the next decade robots should improve to the level of reptiles and within 50 years to the level of mammals. One of the crucial impediments to modeling adaptive behavior in robots, besides the lack of modeling techniques, has been the size and speed of computers. Employing evolutionary techniques for developing neural network based controllers is computationally expensive. Significant progress has been made recently partly due to continuing exponential increase in the computational resources. As Moravec points out, the amount of computational power that a dollar can purchase has increased a thousand-fold every two decades since the beginning of the century. There has been a trillion-fold decline in the cost of computation. If this trend continues, as seems to be the case at present, Moravec predicts that the computational power required for a human-like robot would be available in a $10 million super computer before 2010 and in a $1000 personal computer by the year 2030.

Work on a human-like robot has already begun at MIT with the Cog project [40]. Their approach is to build a humanoid robot that develops and acts in the real world in the same way that humans develop and act. This human-inspired development plan has so far led to the incorporation of several behaviors: the arms have grasping, withdrawal, and reflexes like those of a child; the arms also have adaptive spring-like behavior; the arms follow smooth motion trajectories; the eyes have foveation behavior that can be used to coordinate hand-eye movements in reaching for objects; and the eyes and the head exhibit saccading motion and gaze [41], [42]. Even though Cog's performance today is below those of conventional robots, it is expected that the developmental approach will eventually pay off. Most of the models incorporated in Cog are based on biological models.

It has been suggested that the combination of artificial intelligence with evolutionary computation represents one of the most innovative research directions that may lead to the development of efficient, robust, and easy-to-use solutions to complex real-world problems [10]. Given the incredible computational power at hand, it is becoming increasingly attractive to experiment with evolutionary methods in robots. Onboard computers in mobile systems are now powerful enough to run experiments in real-time, but simulations will probably still continue to play a role for some time. It is also becoming feasible to incorporate robotics into school and college curricula [43]. As robots become less expensive and more prevalent we should expect rapid innovations in the future.

References

[1]   Van de Velde, W. (1991), "Toward Learning Robots," *Robotics and Autonomous Systems*, Vol. 8, Nos. 1-2, pp. 1-6.

[2]   Smithers, T. (1993), "On Why Better Robots Make it Harder," *From Animals to Animats: Proceedings of the Second International Conference on Simulation of Adaptive Behavior*, eds. Meyer, J-A., Roitblat, H., and Wilson, S., MIT Press, Cambridge, MA, pp. 64-72.

[3]   Cliff, D., Harvey, I., and Husbands, P. (1993), "Explorations in Evolutionary Robotics," *Adaptive Behavior*, Vol. 2, No. 1, pp. 73-110.

[4]   Reynolds, C. (1994), "Evolution of Corridor Following Behavior in a Noisy World," in *From Animals to Animats: Proceedings of the Third International Conference on Simulation of Adaptive Behavior*, eds. Cliff, D., Husbands, P., Meyer, J-A., and Wilson, S., MIT Press, Cambridge, MA, pp. 402-410.

[5]   Nordin, P. and Banzhaf, W. (1997), "An On-Line Method to Evolve Behavior and to Control a Miniature Robot in Real Time with Genetic Programming," *Adaptive Behavior*, Vol. 5, No. 2, pp. 107-140.

[6]   Ram, A., Boone, G., Arkin, R., and Pearce, M. (1994), "Using Genetic Algorithms to Learn Reactive Control Parameters for Autonomous Robotic Navigation," *Adaptive Behavior*, Vol. 2, No. 3, pp. 277-305.

[7]   Colombetti, M., Dorigo, M. (1997), "Behavior Analysis and Training-A Methodology for Behavior Engineering," *IEEE Transactions on Systems, Man, and Cybernetics-Part B: Cybernetics*, Vol. 26, No. 3, pp. 365-380.

[8]   Grefenstette, J. (1992), "The Evolution of Strategies for Multi-Agent Environments," *Adaptive Behavior*, Vol. 1, No. 1, pp. 65-90.

[9]   Nolfi, S., Floreano, D., Miglino, O., and Mondada, F. (1994), "How to Evolve Autonomous Robots: Different Approaches in Evolutionary Robotics," *Artificial Live IV: Proceedings of the Fourth International Workshop on the Synthesis and Simulation of Living Systems*, eds. Brooks, R. and Maes, P., MIT Press, Cambridge, MA, pp. 190-197.

[10] Back, T., Hammel, U., and Schwefel, H. (1997), "Evolutionary Computation: Comments on the History and Current State," *IEEE Transactions on Evolutionary Computation*, Vol. 1, No. 1, pp. 3-17.

[11] Holland, J. (1975), *Adaptation in Natural and Artificial Systems*, University of Michigan Press, Ann Arbor, MI.

[12] Mitchell, M. (1996), *An Introduction to Genetic Algorithms*, MIT Press, Cambridge, MA.

[13] Goldberg, D. (1989), *Genetic Algorithms in Search Optimization, and Machine Learning*, Addison-Wesley Publishing Company, New York.

[14] Koza, J. (1992), *Genetic Programming: On the Programming of Computers by Means of Natural Selection*, MIT Press, Cambridge, MA.

[15] Rumelhart, D., Hinton, G., and Williams, R. (1986), "Learning Internal Representations by Error Propagation," *Parallel Distributed Processing*, Vol. 1, eds. McClelland, J. and Rumelhart, D., MIT Press, Cambridge, MA, pp. 318-362.

[16] Elman, J. (1990), "Finding Structure in Time," *Cognitive Science*, Vol. 14, pp. 179-212.

[17] Mondada, R. Franzi, E., and Ienne, P. (1993), "Mobile Robot Miniaturization: A Tool for Investigation in Control Algorithms," *Proceedings of the Third International Symposium on Experimental Robots*, Kytoto, Japan.

[18] Pomerleau, D. (1993), *Neural Network Perception for Mobile Robot Guidance*, Kluwer, Norwell, MA.

[19] Jochem, T. and Pomerleau, D. (1996), "Life in the Fast Lane: The Evolution of an Adaptive Vehicle Control System," *AI Magazine*, Vol. 17, No. 2, pp. 11-50.

[20] Floreano, D. and Mondada, F. (1996), "Evolution of Homing Navigation in a Real Mobile Robot," *IEEE Transactions of Systems, Man, and Cybernetics-Part B: Cybernetics*, Vol. 26, No. 3., pp. 396-407.

[21] Nolfi, S. (1997), "Using Emergent Modularity to Develop Control Systems for Mobile Robots," *Adaptive Behavior*, Vol. 5, No. 3/4, pp. 343-363.

[22] Harvey, I., Husbands, P., and Cliff, D. (1994), "Seeing the Light: Artificial Evolution, Real Vision," *From Animals to Animats: Proceedings of the Third International Conference on Simulation of Adaptive Behavior*, eds. Cliff, D., Husbands, P., Meyer, J-A., and Wilson, S., MIT Press, Cambridge, MA, pp. 392-401.

[23] Harvey, I. (1993), "Evolutionary Robotics and SAGA: The Case for Hill Crawling and Tournament Selection," *Artificial Life III*, ed. Langton, C., Addison Wesley, Reading, MA, pp. 299-326.

[24] Gomez, F. and Miikkulainen, R. (1997), "Incremental Evolution of Complex General Behavior," *Adaptive Behavior*, Vol. 5, No. 3/4, pp. 317-342.

[25] Meeden, L. (1996), "An Incremental Approach to Developing Intelligent Neural Network Controllers for Robots," *IEEE Transactions on Systems, Man, and Cybernetics-Part B: Cybernetics*, Vol. 26, No. 3, pp. 474-485.

[26] Baluja, S. (1996), "Evolution of an Artificial Neural Network Based Autonomous Land Vehicle Controller," *IEEE Transactions on Systems, Man, and Cybernetics-Part B: Cybernetics*, Vol. 26, No. 3, pp. 450-463.

[27] Floreano, D. and Mondada, F. (1994), "Automatic Creation of an Autonomous Agent: Genetic Evolution of a Neural-Network Driven Robot," *From Animals to Animats: Proceedings of the Third International Conference on Simulation of Adaptive Behavior*, eds. Cliff, D., Husbands, P., Meyer, J-A., and Wilson, S., MIT Press, Cambridge, MA, pp. 421-430.

[28] Brady, M. and Hu, H. (1997), "Software and Hardware Architectures of Advanced Mobile Robots for Manufacturing," *Journal of Experimental and Theoretical Artificial Intelligence*, Vol. 9, pp. 257-276.

[29] Jakobi, N., Husbands, P. and Harvey, I. (1995), "Noise and the Reality Gap: The Use of Simulation in Evolutionary Robotics," *Advances in Artificial Life: Proceedings of the Third European Conference on Artificial Life, Granada, Spain, June4-6, 1995*, eds. Moran, F., Moreno, A., Merelo, J., and Chacon, P., Springer-Verlag, Lecture Notes in Artificial Intelligence 929, pp. 704-720.

[30] Husbands, P., Harvey, I., Cliff, D., and Miller, G. (1997), "Artificial Evolution: A New Path for Artificial Intelligence?," *Brain and Cognition*, Vol. 34, pp. 130-159.

[31] Hexmoor, H., Kortenkamp, D., and Horswill, I. (1997), "Software Architectures for Hardware Agents," *Journal of Experimental and Theoretical Artificial Intelligence*, Vol. 9, pp. 147-156.

[32] Harvey, I., Husbands, P., and Cliff, D. (1993), "Issues in Evolutionary Robotics," *From Animals to Animats: Proceedings of the Second International Conference on Simulation of Adaptive Behavior*, eds. Meyer, J-A., Roitblat, H. and Wilson, S., MIT Press, Cambridge, MA, pp. 364-373.

[33] Meeden, L., McGraw, G., and Blank, D. (1993), "Emergent Control and Planning in an Autonomous Vehicle," *Proceedings of the Fifteenth Annual Meeting of the Cognitive Science Society*, Lawrence Earlbaum Associates, Hillsdale, NJ, pp. 735-740

[34] Dorigo, M. and Colombetti, M. (1997), "Precis of Robot Shaping: An Experiment in Behavior Engineering," *Adaptive Behavior*, Vol. 5, No. 3/4, pp. 391-405.

[35] Hexmoor, H. and Meeden, L. (1997), "Learning in Autonomous Robots: A Summary of the 1996 Robolearn Workshop," *Knowledge Engineering Review*, Vol. 11, No. 1.

[36] Grefenstette, J., and Schultz, A. (1994), "An Evolutionary Approach to Learning in Robots," *Naval Research Laboratory Technical Report AIC-94-014*.

[37] Grefenstette, J., Ramsey, C., and Schultz, A. (1990), "Learning Sequential Decision Rules Using Simulation Models and Competition," *Machine Learning*, Vol. 5, No. 4, pp. 355-381.

[38] Clark, A. (1993), *Associative Engines: Connectionism, Concepts, and Representational Change*, MIT Press, Cambridge, MA.

[39] Moravec, H. (1992), "The Universal Robot," *Visions of the Future: Art, Technology and Computing in the Twenty-First Century*, ed. Pickover, C., St. Martin's Press, New York, pp. 65-73.

[40] Brooks, R. and Stein, L. (1994), "Building Brains for Bodies," *Autonomous Robots*, Vol. 1, No. 1, pp. 7-25.

[41] Ferrell, C. (1996), "Orientation Behavior Using Registered Topographic Maps," *From Animals to Animats: Proceedings of the Fifth International Conference on Simulation of Adaptive Behavior*, MIT Press, Cambridge, MA.

[42] Marjanovic, M. Scassellati, B., and Williamson, M. (1996), "Self-Taught Visually Guided Pointing for a Humanoid Robot," *From Animals to Animats: Proceedings of the Fifth International Conference on Simulation of Adaptive Behavior*, MIT Press, Cambridge, MA.

[43] Meeden, L. (1996), "Using Robotics as an Introduction to Computer Science," *Proceedings of the Ninth Florida Artificial Intelligence Research Symposium*, ed. Stewman, J., pp. 473-477.

INDEX